Statistical Methods and the Geographer

GEOGRAPHIES FOR ADVANCED STUDY

Edited by Professor Stanley H. Beaver, M.A., F.R.G.S.

THE BRITISH ISLES: A GEOGRAPHIC AND ECONOMIC SURVEY
CENTRAL EUROPE
CONCEPTS IN CLIMATOLOGY
EASTERN EUROPE
GEOGRAPHY OF POPULATION
GEOMORPHOLOGY
THE GLACIATIONS OF WALES AND ADJOINING REGIONS
AN HISTORICAL GEOGRAPHY OF SOUTH AFRICA
AN HISTORICAL GEOGRAPHY OF WESTERN EUROPE BEFORE 1800
HUMAN GEOGRAPHY
INTRODUCTION TO CLIMATIC GEOMORPHOLOGY
LANDFORMS OF THE HUMID TROPICS, FORESTS AND SAVANNAS
LAND, PEOPLE AND ECONOMY IN MALAYA
MALAYA, INDONESIA, BORNEO AND THE PHILIPPINES
NORTH AMERICA
THE POLAR WORLD
A REGIONAL GEOGRAPHY OF WESTERN EUROPE
THE SCANDINAVIAN WORLD
THE SOVIET UNION
STATISTICAL METHODS AND THE GEOGRAPHER
THE TROPICAL WORLD
URBAN ESSAYS: STUDIES IN THE GEOGRAPHY OF WALES
URBAN GEOGRAPHY
WEST AFRICA
THE WESTERN MEDITERRANEAN WORLD

Statistical Methods and the Geographer

S. GREGORY
Professor of Geography, University of Sheffield

Third Edition

LONGMAN

LONGMAN GROUP LIMITED
London

Associated companies, branches and representatives throughout the world

First published 1963
Second Edition 1968
Third Edition 1973
Second impression 1975

ISBN 0 582 48423 5 paper

PRINTED IN GREAT BRITAN BY
LOWE AND BRYDONE (PRINTERS) LTD., THETFORD, NORFOLK

To my wife, Marjorie
whose devotion and support have never failed

PREFACE

The origins of this book lie in the author's experiences, as student, research worker and lecturer, over the past 15 years. The intricacies and essential characteristics of statistical methods were first introduced to him as a student by Professor P. R. Crowe, when the latter was a Reader in the University of London. The value of such methods of analysis has been increasingly appreciated as research, especially in the field of climatology, has been pursued during succeeding years.

For the non-mathematician, however, even the simpler introductory books on statistics often raise considerable problems. These are accentuated, moreover, by the fact that the methods are applied to fields of study which are, in large measure at least, unfamiliar to the geographer—industrial or business control, sociology or economic theory, the biological sciences or medicine, or simply as a study in applied mathematics. Moreover, most geographical studies that have employed statistical techniques have equally tended simply to assume that the reader would understand the methods despite the normal lack of formal statistical training.

In an attempt to counteract these tendencies, training in statistical methods for geography students was expanded at Liverpool University in 1957. This training aimed at providing a grounding in a variety of basic methods, all of which were developed and applied in terms of geographical problems. From the course has evolved the present book, which it is hoped will provide a similar basic grounding for all geographers. Throughout the evolution of this course, and especially in encouraging me to expand it in the present form, I have had every support from Professor R. W. Steel. It is my former colleague, Dr A. T. A. Learmonth, however (now Professor of Geography in the School of General Studies, Australian National University, Canberra), to whom the greatest debt is owed, for his unfailing willingness to discuss and constructively criticize my efforts, for his persistence in exhorting me to proceed with the work, and for invaluable advice and assistance.

There are many others who, in their various ways, have provided help and guidance. Amongst these are Professor S. H. Beaver, who read and commented on the text; Dr D. J. Bartholomew, formerly lecturer in Statistics at the University of Keele, whose advice at an early stage helped to set the pattern of this book; Mr P. K. Mitchell, the Geography Department, the University College of Sierra Leone, a colleague during my year in Sierra Leone (1960–1961), when the bulk of this book was written; Miss P. J. Treasure of the Geography Department, the University of Liverpool, who drew most of the diagrams; Miss E. M. Shaw, the University of Keele, for help at proof stage; and all my colleagues at Liverpool who willingly allowed me to try my ideas upon them.

To them, and to many others, my thanks are due—I trust that they approve of the final product.

S. G.

LIVERPOOL, 1962

SECOND EDITION

The continually growing interest amongst geographers, at research, undergraduate and now sixth-form levels, in the relevance of statistical techniques to the subject, has made it desirable that this book should undergo some revision, both to meet wider demands and to satisfy the need for more effective presentation. The author is extremely grateful to those many friends and colleagues who wrote to him concerning the first edition, whether this was to commend sections which seemed satisfactory or to raise questions concerning others which did not completely fall into this category. Their comments and advice have all been seriously considered when the text was being revised, even when action has nevertheless not been taken.

Apart from innumerable minor textual changes which, it is hoped, will make for more effective reading and understanding, the major changes are fivefold. First, some attention has been given to the problem of the transformation of data in order to reinforce the appreciation of the need for normally-distributed data for the use of so many techniques. Secondly, the use of probability paper, at least in simple terms, has been introduced to illustrate the ways in which the labour of probability assessments can be circumvented. Thirdly, radical changes have been made, plus considerable expansion added, to the theme of non-parametric testing, to provide a more systematic approach to what is a most important group of possible techniques for geographers. Fourthly, change and expansion are also reflected in the sections on correlation and regression, including some simple consideration of curvilinear relationships and the presentation of computational techniques more geared to the use of desk calculators rather than long-hand methods. Fifthly, the bibliography has also been expanded, to incorporate a wider range of books on techniques and a selection of research papers using such techniques in a geographical (or near-geographical) context.

Nevertheless, the overall structure and framework of the book remains basically unaltered; the intellectual approach is still one of presenting the techniques as simply as possible, leaving those requiring more advanced techniques to move on to the appropriate advanced texts, and of allowing geographers to be introduced to these potentially valuable methods through examples which, it is hoped, seem relevant to them, i.e. in terms of geographical problems. The growth of the use of such methods, both of the simple and the complex varieties, over the past few years not only further stresses the need for all geographers to be conversant with such techniques and their implications, but also encourages the author to hope that these improvements in this introductory text will be appreciated and approved of by all its users.

S. G.

LIVERPOOL, 1967

THIRD EDITION

The need to modify a book of this sort is always apparent and this opportunity to do so has been gratefully appreciated. Fundamental changes are few, however, and the express purpose of providing a simple, largely non-technical introduction, for non-numerate geographers, has been retained.

In the four years since the second edition was prepared, the number of books on more advanced techniques, and of articles and books using such techniques in geography, has expanded considerably, and many of these are included in the bibliography. The types of problems studied by geographers have also become more varied, but changes to incorporate some of these newer fields have deliberately not been made, for the methods presented are independent of specific fields and approaches. Moreover, the more traditional and simpler problems used here are perhaps more suitable as basic teaching examples—other uses and applications can be found in the texts listed.

This continued growth of more advanced work, itself a most welcome development, implies that a growing number of practising geographers are now operating at levels far beyond those in this book. It is still necessary, however, for the initial stages of statistical manipulation and thinking to be learned and appreciated by all those entering the field. It is hoped that, for these, this book will continue to prove of value and help as they take their first steps along this rocky but rewarding path.

S. G.

SHEFFIELD, 1972

CONTENTS

ACKNOWLEDGEMENTS

We are grateful to the Syndics of the University Press, Cambridge, for permission to include adaptations of tables from *Cambridge Elementary Statistical Tables* by D. V. Lindley and J. C. P. Miller. Table 23 was reproduced from Table L which appeared in *Non-parametric Statistics for the Behavioral Sciences* by S. Siegel (McGraw-Hill Book Company), and was abridged from the sources: *Annals of Mathemetic Statistics* and *Psychological Bulletin*, and we are indebted to the McGraw-Hill Book Company, the Treasurer of the Institute of Mathematical Statistics and the Managing Editor of American Psychological Association for their respective permissions for the inclusion of this copyright material.

INTRODUCTION

The type of geography which admits the importance of quantification and the appropriateness of statistical methodology, but always as servants and not as masters, would appear to be the best answer the profession can furnish to the embarrassing questions which have arisen during the current debate in academic circles regarding geography's right to be included in the curricula of institutions of higher learning.

WILLIAM WARNTZ

In this third quarter of the twentieth century the raw material with which the geographer deals has become progressively more of a quantitative nature and less merely qualitative. This gradual but steady change in emphasis has of necessity engendered a modification of the intellectual approach to the subject. As in any other worthwhile field of study, so in geography each generation attempts to absorb, and then advance beyond, the accumulated work of previous generations; this is no more than the outward sign of healthy development. These advances may at times be in terms of factual knowledge. At other times, however, they reflect a changing approach to the subject at large, such as this present conscious and deliberate attempt to provide a more quantitative approach to the geographer's problems.

In all branches of the subject this tendency has developed. Climatological investigations have traditionally and necessarily been concerned with numerical data. Economic geography, too, has for long utilized quantitative data as a prime source of information, although explanatory studies have tended to rest more heavily on subjective judgments than would in many cases seem desirable. Geomorphology, population studies and various other aspects of human geography, amongst many branches of the subject, have also increasingly turned to more precise numerical data over the recent past, all in the attempt to render a more accurate and objective assessment of the geography of particular areas or problems. Moreover, as geographers increasingly co-operate with scientists from other disciplines, or engage in the practical fields of planning, the need to present both data and conclusions in sound quantitative terms becomes even more pressing.

Once such an attitude is accepted, however, a necessary corollary follows, that these numerical data should be analysed by sound statistical methods so that maximum value is obtained from them. Too often a considerable body of valuable quantitative data is presented either in a raw state or after a minimal amount of processing. Sometimes, of course, this may be quite legitimate as it is all that the problem requires. In other cases, however, more fundamental, and possibly more valid, conclusions could be reached, or varied aspects of a problem investigated, by means of a more comprehensive and subtle use of existing statistical methods. Moreover, it is not simply that such methods are not always used, but that at times false interpretations are made either because of the failure to apply such methods or because they are misunderstood. The latter may unfortunately arise when a geographer quite properly consults a professional statistician without at the same time fully understanding the implications of the results which are obtained by the methods with which he is provided.

The aim of this book is therefore to present standard statistical techniques in a simple manner and to apply them to problems typical of those which geographers consider. In this way a twofold purpose is served. On the one hand the requirements of practising geographers engaged in research are at least partially met by the presentation of methods and techniques, at a relatively simple level, which should enable many geographical problems to be analysed more soundly. This is not intended to be a comprehensive work covering the full field of statistics, but rather a selective presentation of elementary methods, which are particularly applicable to geographical problems. For the investigation of more complex problems the standard statistical texts, of which a selected list is incorporated in the bibliography, must be consulted. On the other hand, the introduction to relevant elementary statistics which this book will provide will enable all students of geography more readily to interpret and understand studies based on statistical analyses. Many of the misinterpretations which occur at present result from the *reader's* failure to be conversant with either the advantages or the limitations implicit in any writer's statistical methods—this renders difficult the full and accurate assessment of the value and implication of what is written. From both viewpoints, therefore—from that of the geographer trying to analyse and present his material more effectively, and that of the student of geography trying to interpret and understand existing studies—it is hoped that this excursion into statistical methods and their uses to geographers will prove of value.

A fundamental difficulty arises here, however, and it is one which is inherent in the whole training which most potential geographers receive from their childhood onwards. Most aspiring geographers, though by no means all of them, have indeed studied mathematics to Ordinary level of the G.C.E. In far too many schools, however, it is either administratively impossible, or academically not permissible, to study both geography and mathematics together up to Advanced level. This lack of sixth-form training, or perhaps the actual training received prior to that date, tends to leave many prospective geographers with a built-in resistance to anything which vaguely suggests mathematics. Directly $(a + b)$ is written on the blackboard, or a square root is required, a mental barrier is irrationally erected. This quite needless refusal to attempt to tackle such problems tends to nullify attempts to put geography on a sounder footing in its handling of quantitative data.

Throughout this book, therefore, the deliberate design is to lead the reader by the hand through these apparently difficult by-ways. Save where it is absolutely necessary, there is no attempt to delve into the mathematical theories behind the methods, but rather the concepts involved are presented in plain English instead of, or as well as, in symbols. The computational problems involved should not unduly strain the capabilities of any normally intelligent fifteen-year-old. What is required, on the other hand, is a conscious willingness to follow a statistical argument through to its logical conclusion, to breach this mental barrier of which I have written and in that way to discover an invaluable tool which has been neglected by geographers for far too long. A short selection of publications which have used these techniques is included in the bibliography.

Thus this book is not designed for statisticians; nor does it claim to make statisticians of those who work their way through it. Many possible methods, or applications of methods, which could have been included have instead been deliberately omitted. Rather, a selection of useful methods that can be applied in the field of geography are presented, and illustrated in terms of problems which the geographer can understand. It is only in this application in terms of geography that the author can claim to have made any personal contribution, for the methods presented are in common use in so many other disciplines already, and explained—in greater or lesser complexity and clarity—in numerous other books. It is to this wide range of statistical texts at a more advanced level, such as those included in the bibliography, that the enthusiast or the specialist must turn, if the series of simple illustrations in this book stimulates him to further enquiry. It is not primarily as an introduction to these more advanced statistical studies that this book is designed, however. If, instead, it succeeds in enabling geographical students to handle and interpret quantitative data more effectively, then the author will feel that it has more than fulfilled its purpose.

CHAPTER 1

THE NATURE OF THE RAW MATERIAL

The methods and techniques used in the analysis of statistical data are in large measure controlled by the very character of the statistical data themselves. It is therefore necessary to begin with a very brief consideration of some of these characteristics so that the varied themes that will be introduced later will be more readily understood.

When any collection of figures, representing some quantitative value of any given phenomenon, is to be processed it will be found that although such figures all represent the same phenomenon they are not all of exactly the same value. Thus if a study were being made of the distance inland from the coast that vessels of a given draught could sail it would be found that these distances vary markedly between one river and another, or between one part of the world and another. Again, if the number of vessels sailing along these rivers were examined a very wide range in values would be found between the different rivers. This highly variable nature of the numerical data is common, to a greater or lesser extent, to all sets of data, and this quantity which varies (mileage, or numbers of vessels, in the two cases given above) is known as the *variate.*

A fundamental distinction must be made between two different types of variate, however. In the case of the navigable mileage of rivers outlined above, it is possible for *any* mileage value to be recorded and for fractions of a mile to be included. In other words, it is a *continuous* variate such that there are no clear-cut or sharp breaks between the values that are possible. On the other hand, the number of vessels actually sailing these rivers can only be in terms of *whole* numbers or integers, and fractions of a vessel cannot be recorded. Such a variate is known as *discrete*, and special care must be taken when basing conclusions on the analysis of such discrete variates, as will be seen later.

This variable nature of conditions can best be understood and appreciated if the data are plotted graphically to show the frequency of occurrence of values of different given amounts. The data are first grouped into 'classes', so that it is known how many occurrences fall into each of a series of quantitatively different sets of conditions.

Then the number of occurrences are plotted against the appropriate 'class', and a diagram drawn in the form of 'building blocks'. Such a diagram is known as a histogram and the pattern which it presents is called the frequency distribution for that set of data. From such a diagram a smoothed curve can be interpolated, this being known as the 'frequency curve' of that set of data. Thus in Fig. 1 can be seen the frequency distribution for population densities of the European nation-states. The values for individual states are grouped into various classes depending on their order of magnitude (e.g. 0–49·9

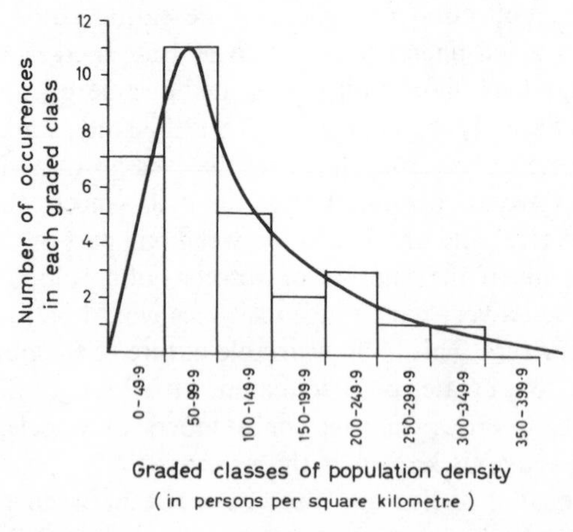

Figure 1. Histogram and frequency distribution curve for population densities of the European nation-states

persons per sq. km.; 50–99·9 persons per sq. km.), and the variable character of these population densities is readily apparent. The way in which these population densities vary is shown by both the 'blocks' and by the smoothed curve. A similar frequency distribution curve can be constructed for any and all sets of data. Fig. 2, for example, shows the distribution of hill summit heights in North Wales based on summit ring-contours taken from the O.S. 1 : 25,000 maps. As with the population densities, these summit heights are a continuous variate. Moreover, both Fig. 1 and Fig. 2 also display another feature of many distribution curves. It can be seen clearly that these curves are not symmetrical, having their peak markedly to one side. Such a distribution is known as *skew*, and the problems which this introduces, together with various methods by which these problems may be largely solved, will be considered later.

It is, in fact, mainly because of the variable character of sets of data, as well as the fact that the distribution curves which reflect this

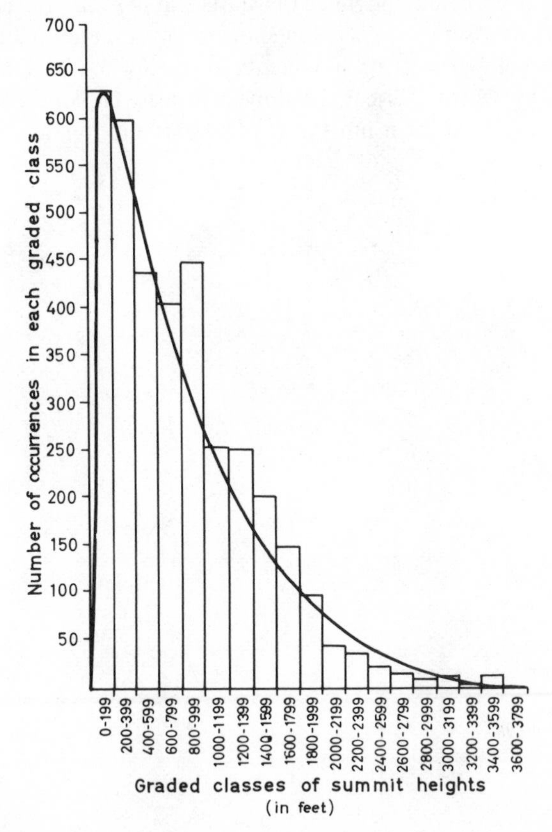

Figure 2. Histogram and frequency distribution curve for hill-summit heights in North Wales

variability tend to differ from each other in terms of skewness, that the whole need for sound statistical analysis of numerical geographical data arises. If values for any given phenomenon were always the same most of the analyses would be unnecessary, for direct comparison of these unvarying values would usually be adequate. Another reason why careful analysis is required, however, is that very often it is not possible to obtain data for the whole of the conditions with which one is concerned. Rather it is a matter of considering a *sample* of these conditions, working on the assumption that this sample provides a fair representation of the whole body of data (the latter being known as the statistical *population*). The extent to which this assumption is justified or not must therefore be allowed for when comparisons or judgments are being made. The various methods by which this can be done will also be considered later, although the need for it must always be borne in mind.

CHAPTER 2

THE CALCULATION AND USE OF THE MEAN

The previous chapter has shown that sets of data are usually composed of individual values which vary from one another to a greater or lesser extent. When, however, it is necessary to express the quantitative aspect of such a set of data briefly and succinctly, i.e. to employ *descriptive statistics*, a lengthy recital of all the individual values is not of much use. Even the graphical representation of these, as illustrated in Chapter 1, is not a great help, for it neither allows of a speedy and easy comparison between different sets of data nor of a ready expression of these characteristics in words or numbers. It is therefore often very useful to be able to summarize these varying values within the one set of data by *one* value alone. This one value is chosen so as to give as reasonable an approximation as possible to what is 'normal' or, in other words, to summarize the data by some 'measure of central tendency'. It is immediately apparent that however this measure is chosen it must involve certain generalizations and must also obscure many characteristics of the set of data that the distribution curve shows.

Types of Mean

Such a generalized summary of conditions can be obtained in various ways, and a few simple illustrations will show the differences between them. If a set of data were

1, 2, 3, 4, 5

and it were necessary to summarize these values, most people would carry out such a summary in the following way. They would probably add the numbers together, getting a total of 15, and then divide this by the total number of items, i.e. by 5. In this way they would arrive at an answer of 3. On the other hand it would also be possible to arrange the values in order of magnitude—as they are already arranged above—and choose the middle one as being representative of them all. Again, the answer would be 3.

With a rather more complex set of data the same approach could be adopted. For example, the data may be as follows:

1, 2, 2, 3, 3, 3, 3, 4, 4, 5

In this case the total of these several values is 30 and if this is divided by the number of values involved, i.e. 10, the answer is again 3. Also, if the method of choosing the central value when they are arranged in order of magnitude is used, then 3 is again the answer. In this case, another method also presents itself, for it is possible to proceed as for the preparation of a distribution curve and group the data into sets or classes in the following way.

Value	Number of Occurrences
1	1
2	2
3	4
4	2
5	1

Having thus grouped the data according to the number of occurrences of any one value it is possible to choose that value which occurs most frequently. Here it is once again the value 3.

It is these three methods, presented here in all their simplicity, that are the three basic ways of summarizing a set of data. Each of these methods gives a value which to some extent represents the set of data. This value is sometimes referred to as the 'normal' or 'norm', but the more usual term for it is the 'mean value' or, more simply, the 'mean'.

The first method applied above is the *arithmetic average* or the *arithmetic mean*, usually more loosely referred to as either the *mean* or the *average*. As was seen when the method was first used, the average is simply obtained by adding the values together and then dividing by the number of values that there are. It can be defined more precisely—and apparently more technically—by stating that 'the average is a quotient obtained by dividing the total by the number of occurrences connected with it'.

To express this relationship in mathematical terms is quite simple once the 'shorthand' used is memorized. Thus the above statement can be written as:

$$\bar{x} = \frac{\sum_{i=1}^{n} x}{n}$$

In this statement

x = the individual values making up the series of data

$\bar{x}$ = the average of that series

n = the number of occurrences being considered

$\sum_{i=1}^{n}$ = the summation of all the values of x, from the first ($i = 1$) to the final one (n), i.e. the total when all values of x are added together. In later pages, the subscripts and superscripts ($i = 1$ and n, or their equivalents) are omitted unless they are essential, summation being indicated by $\sum$ alone.

Thus in a few signs the rather long definition of the average given above can be summarized easily.

The second mean value which was set out in the earlier examples is called the *median*. This was obtained by placing the values in ascending or descending order of magnitude or rank and then finding the central value of these. If the total number of occurrences were to be an odd number then the median would be one of the observed values. This can be generalized by stating that the appropriate point in the ranking scale is $\frac{n+1}{2}$ when n is the number of occurrences or ranks. If, on the other hand, there were an even number of occurrences, then the median would lie midway between two of these values. These differing sets of conditions are to be seen in the two simple examples with which this consideration was begun. In the first case $n = 5$, so that the median rank $= \frac{5+1}{2} = \frac{6}{2} = 3$, i.e. the third ranked value. In the second case $n = 10$, so that the median rank $= \frac{10+1}{2} = \frac{11}{2} = 5{\cdot}5$, i.e. midway between the fifth and sixth ranked values. Once again, however, it is as well to define terms as precisely and unambiguously as possible and to state that 'the median is the reading on the scale of the variable such that there are an equal number of entries above it and below it'.

The third mean which was considered, in which the data were grouped into classes and the class containing the most occurrences was chosen, is referred to as the *mode*, i.e. the most fashionable; the one which occurs most often. This again can be presented in terms of a formal definition, such as that 'the mode is the value of the class within a statistical group in which there are most incidences'.

Relationships between the Means

These three—the average, the median and the mode—are the main methods of expressing the mean value of any set of data. It would therefore seem desirable to consider briefly the relationship between them and to try to assess which, if any, is preferable to the others, and why this may be so. This can first be done by considering one of the earlier examples in a slightly different way. If the series

1, 2, 2, 3, 3, 3, 3, 4, 4, 5

were to be plotted as a histogram it would appear as in Fig. 3, where a smooth frequency distribution curve is also drawn to fit these data. In this particular example, as has been seen above, the three mean values all coincide at the same point, i.e. 3. Also the accompanying frequency distribution curve is perfectly evenly balanced on either side of these mean values. This perfect coincidence of the three mean values, plus the symmetrical distribution curve, are characteristic of what is termed a *normal* frequency distribution. Other characteristics of the normal curve are considered in Chapter 4, where it will be seen that the distribution curve in Fig. 3 is not strictly normal —for now, however, we can perhaps accept it as approximating to normal. The existence of such a normal distribution is assumed in most statistical methods, although in practice it is seldom perfectly achieved, as was indicated in Chapter 1.

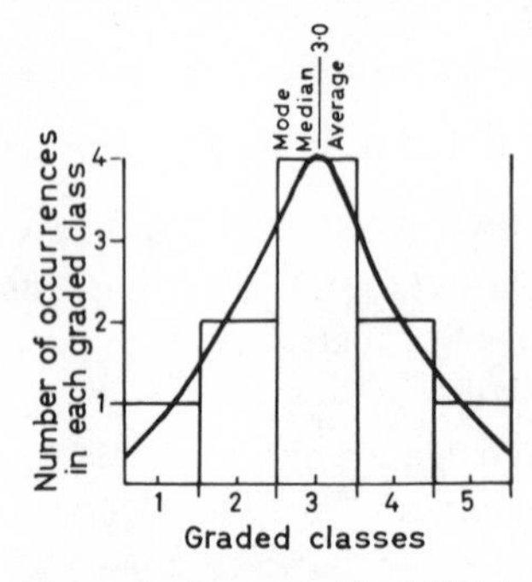

Figure 3. A symmetrical distribution curve that approximates to normal

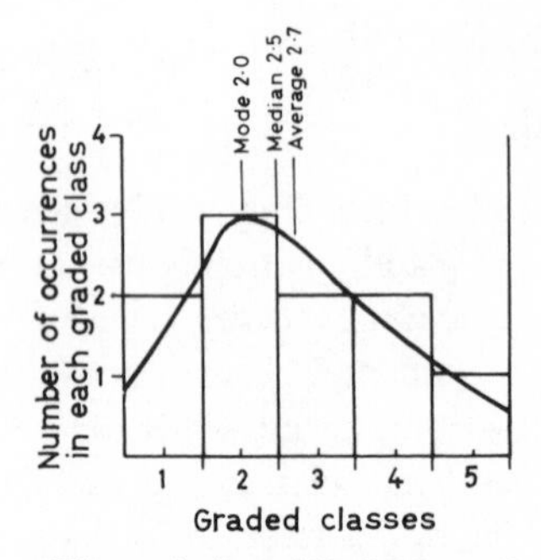

Figure 4. A positively skew distribution curve

2 THE CALCULATION AND USE OF THE MEAN

The following set of data illustrates a non-normal distribution:

1, 1, 2, 2, 2, 3, 3, 4, 4, 5

The three means can be calculated as below:

(*a*) the average $= \bar{x} = \frac{\Sigma x}{n} = \frac{27}{10} = 2{\cdot}7$

For the median and the mode the values can be retabulated

values:	1, 1,	2, 2, 2,	3, 3,	4, 4,	5
occurrences:	2	3	2	2	1

(*b*) median 2·5
(*c*) mode 2

Thus, in this case, the three means are different from each other, the average being the largest value and the mode the smallest. In Fig. 4 these values are plotted on a histogram and the distribution curve is added. Clearly this distribution curve is NOT evenly balanced, having its 'peak' to the left of centre and a 'tail' to the right. This lack of balance is called *skewness*, while when the 'tail' extends to the right, as in this case, it is classed as *positive* skewness.

On this diagram are also entered the three means, the mode lying to the left, the median in the centre and the average to the right. This relative pattern of the three means is true of all positively skew distributions. Moreover, provided that the skewness is not too marked, a general quantitative relationship also tends to exist between the means.

This relationship, which gives only a broad approximation, can be expressed as follows:

MODE = AVERAGE − 3(AVERAGE − MEDIAN)
i.e. MODE = 2·7 − 3(2·7 − 2·5)
= 2·7 − (3 × 0·2)
= 2·7 − 0·6 = 2·1

In fact the modal value is 2·0. Expressed in other terms, it means that the median lies one-third of the way back from the average towards the mode (Fig. 5).

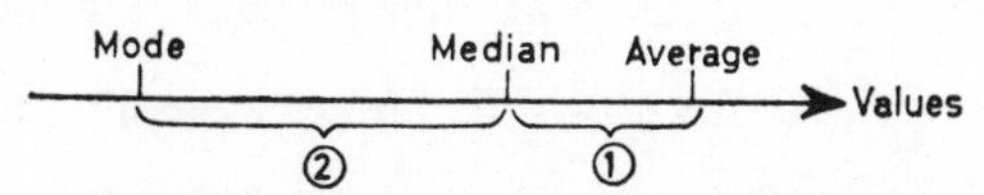

Figure 5. Relations between the means in a skew distribution

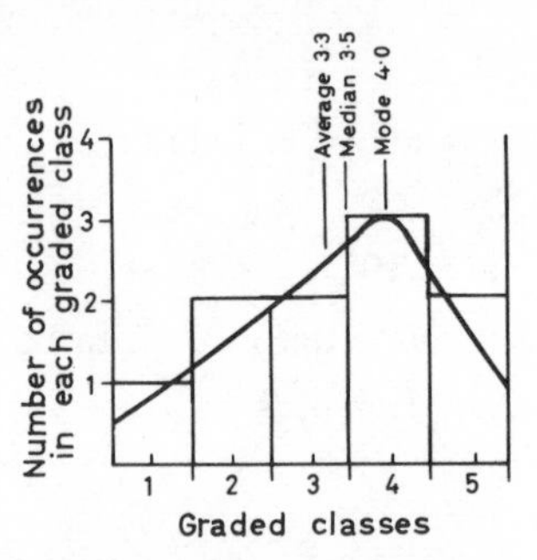

Figure 6. A negatively skew distribution curve

It is equally possible for distributions to be *negatively* skew, i.e. for the 'tail' to lie to the left of the curve and for the peak to lie to the right. This is exemplified in the following case (Fig. 6):

values: 1, 2, 2, 3, 3, 4, 4, 4, 5, 5
occurrences: 1 2 2 3 2

$$\text{average} = \frac{33}{10} = 3{\cdot}3 \quad \text{median} = 3{\cdot}5 \quad \text{mode} = 4{\cdot}0$$

The general relationship of the mode, the median and the average still holds true but in the reverse direction,
i.e. MODE = AVERAGE + 3(MEDIAN − AVERAGE).

Specific Examples

Having outlined these methods in abstract terms they should now be considered in relation to specific data of geographical interest. In Table I are set out the annual totals of rainfall at Bidston Observatory, Birkenhead, for the thirty years 1901–1930. These vary between 22·47″ and 36·50″ and these data represent a continuous variate. On calculating the average it is found that

$$\frac{\Sigma x}{n} = \frac{853{\cdot}63}{30} = 28{\cdot}45''$$

In the second column of Table I these values are retabulated into order of magnitude. As there are thirty values, the median rank $= \frac{30 + 1}{2} = 15{\cdot}5$, so that it will lie between the fifteenth and sixteenth values, i.e. midway between 28·08″ and 28″45″, giving a value

Table I

Annual rainfall at Bidston Observatory, Birkenhead, 1901–1930

Values in order of occurrence	Values in order of magnitude		Classes	No. of occurrences
x (inches)	(inches)		(inches)	
25·19	22·47			
25·57	24·01			
34·42	24·87			
25·18	25·15			
24·01	25·18			
28·08	25·19			
26·57	25·27			
28·90	25·57			
28·45	25·78			
28·59	25·97			
25·27	26·02			
30·17	26·57		21–22·99	1
25·78	26·83		23–24·99	2
26·02	28·00		25–26·99	10
26·83	28·08	Median	27–28·99	6
24·87	28·45	28·27	29–30·99	5
30·59	28·59		31–32·99	2
31·93	28·90		33–34·99	3
29·12	28·95		35–36·99	1
33·34	29·11			
22·47	29·12		Mode = 25–26·99	
25·97	30·17			
30·92	30·59			
32·87	30·92			
28·00	31·93			
28·95	32·87			
34·81	33·34			
29·11	34·42			
25·15	34·81			
36·50	36·50			
30)853·63				
Average 28·45				

of 28·27″. In the third and fourth columns of Table I, the values are grouped into several classes and the number of occurrences in each class is shown. As this is a continuous variate, all possible numerical values must be allowed for, not simply whole numbers. The class limits must therefore be designed to provide a fully continuous range

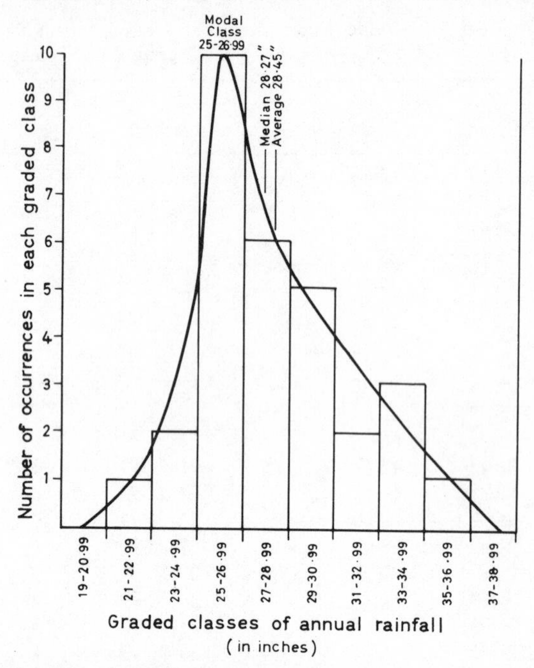

Figure 7. Histogram and frequency distribution curve for annual rainfall at Bidston, 1901–1930

of values, i.e. not 21 and 22; 23 and 24, etc. but 21 to 22·99; 23 to 24·99, etc. In this way a modal class of 25″–26·99″ is found to occur. These conditions are represented graphically in Fig. 7 where slight positive skewness is shown.

It is also possible to calculate a specific *modal value*, using the formula:

$$L + \frac{Fm-Fb}{(Fm-Fb)+(Fm-Fa)} \cdot C,$$

when L = the lower limit of the model class; Fm = the frequency of (or number of items in) the modal class; Fa = the frequency of the class *above* the modal class; Fb = the frequency of the class *below* the modal class; C = the class interval, i.e. the size of the value range of each class.

Thus, from Table I:

$$\text{Modal Value} = 25{\cdot}00 + \frac{10\text{–}2}{(10\text{–}2) + (10\text{–}6)} \cdot 2$$

$$= 25{\cdot}00 + \frac{8}{12} \cdot 2 = 25{\cdot}00 + 1{\cdot}33$$

$$= 26{\cdot}33.$$

Moreover, at times, geographical data can be obtained only in the form of such groups or classes, and, if median or average values are then needed, they can be calculated in an approximate form. The calculation of the average from grouped data is presented more conveniently later on (p. 33 and Table VII), but the median can be calculated from the formula:

$$M = l + \frac{fm}{fc} \cdot \mathrm{C},$$

when M = the median; l = the lower limit of the class in which the median occurs; fm = the number of items between the lower limit of that class and the median rank; fc = the frequency in the class containing the median; C = the class interval.

Thus, in the case of the Bidston data, it has been shown that the median rank is 15·5, and, from the class data in Table I, this rank must fall in the class with limits of 27·00″ and 28·99″, which contains six items, ranked from fourteenth to nineteenth. So:

$l = 27{\cdot}00$ $C = 2{\cdot}00$ $fc = 6$ $fm = 2{\cdot}5$ (i.e. the median rank is 2·5 ranks above the bottom of the class)

$$\text{Therefore, } M = 27{\cdot}00 + \left(\frac{2{\cdot}5}{6} \times 2\right) = 27{\cdot}00 + 0{\cdot}83$$

$$= 27{\cdot}83.$$

This is *not* the same as the previous accurate calculation, the error being larger than usual, because all six items in the class actually fall in the upper half of that class.

Differences in frequency distributions and in the relationship between the three means can also be appreciated by considering data related to economic geography. In Table II are set out the annual iron-ore production figures for the twenty years 1938–1957 for four western European countries—Belgium, France, Luxembourg and the United Kingdom. The average and median values can be readily obtained, in the same way as for the Bidston rainfall data, and these are given at the foot of each column. In every case the median is higher than the average, suggesting a tendency for negative skewness, though in the case of France this does not seem to be borne out by

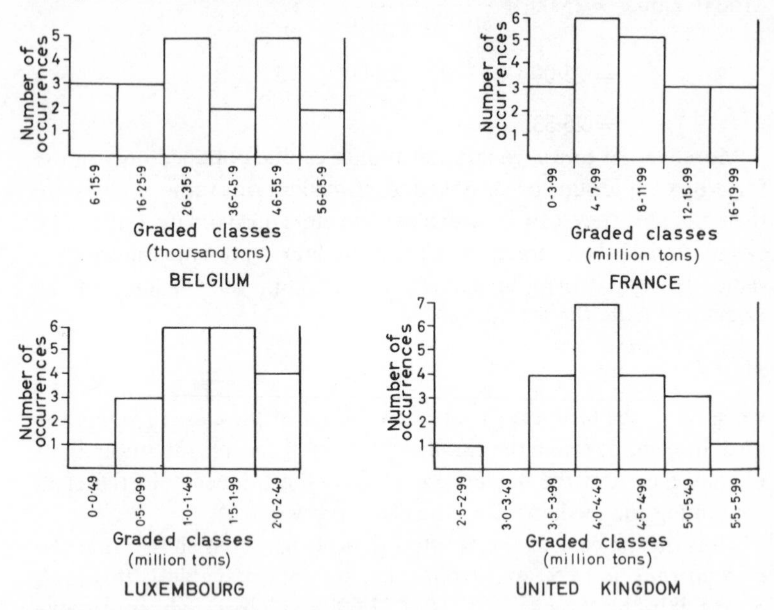

Figure 8. Histograms of annual iron-ore production for Belgium, France, Luxembourg and the United Kingdom, 1938–1957

the frequency distribution (Fig. 8). Fig. 8 also displays the difficulty of establishing a clear-cut modal group in many sets of data.

Advantages and Disadvantages

The advantages and disadvantages which the three types of mean possess as working tools can now be more generally considered, following these illustrations from actual conditions. The mode, by its very definition, indicates that which is most common or frequent. Very often, however, there is some difficulty in deciding exactly where the mode occurs. This difficulty can arise for one of two reasons. First, the distribution may not be *unimodal*, i.e. it may well have two or more modal groups of roughly equal importance. Thus, when considering the iron-ore data (Table II) it was seen that it was difficult to establish a clear mode, especially for Belgium and Luxembourg. In each of these cases two classes of equal frequency

Table II

Annual iron-ore production 1938–1957 (in thousands of tons)

	Belgium	France	Luxembourg	U.K.
	65	10,203	1,506	3,615
	60	10,161	1,639	4,417
	29	4,113	1,368	5,449
	47	3,467	1,912	5,528
	41	4,144	1,431	5,449
	46	5,350	1,471	5,411
	16	2,862	816	4,390
	11	2,349	394	4,162
	14	5,021	650	3,574
	21	6,099	592	2,974
	34	7,555	1,020	3,990
	15	10,200	1,241	4,086
	16	9,750	1,154	3,812
	28	11,440	1,688	4,504
	47	13,230	2,174	4,618
	35	13,790	2,151	4,500
	29	14,240	1,766	4,369
	37	16,340	1,933	4,437
	50	17,120	2,034	4,457
	48	18,770	2,036	4,637
Average	34·45	9,310·2	1,448·8	4,418·95
Median	34·5	9,955·5	1,488·5	4,427·0

exist in all suitable broad groupings of the data (Fig. 8). The second difficulty arises from the selection of the classes which are to be adopted.

If the Bidston rainfall data, for example, were to be grouped in terms of classes beginning 22″–23·99″, instead of 21″–22·99″ as in Table I and Fig. 7, the following frequencies would be found:

Classes (inches)	No. of occurrences
22–23·99	1
24–25·99	9
26–27·99	3
28–29·99	8
30–31·99	4
32–33·99	2
34–35·99	2
36–37·99	1

The modal class would thus become 24″–25·99″ instead of 25″–26·99″, the modal value 25·14″ instead of 26·33″, while the frequency distribution would appear to be virtually duomodal. These difficulties mean that, in practice, the mode is a very imprecise form of the mean value; it may be difficult to locate and the actual value arrived at may in part result from a subjective choice of groupings. Furthermore, the mode does not possess any true mathematical qualities having at best only a generalized relationship to the average (p. 9), so that it cannot be used in formulae to derive further characteristics of the set of data. Save for graphical purposes (and also for its use in certain generalized computations to be outlined later) the mode is not a method to be highly recommended.

When the median is used as the mean value, it can be considered as representing the 'mean expectation', in that there are as many individual occurrences above it as there are below it. Moreover, in the calculation of the median every occurrence is given the same unit weight, i.e. it is regarded as of equal importance, whether it be of small, medium or large magnitude. Indeed, magnitude of individual values is of *no* importance directly, except for that of the central value when there is an odd number of occurrences being considered, or of the central two values when there is an even number of occurrences. This means that widely differing sets of data can return the same median value, as is indicated in a generalized way in Fig. 9. Furthermore, this implies that the median possesses no real mathematical qualities and cannot be used for further computation except in the most general manner. In this way it suffers from the same limitation as does the mode. Nevertheless, the median does possess the valuable property of clear definition, in that its relative position within the occurrences is undisputed and readily understood, while it is also very useful in illustrative material.

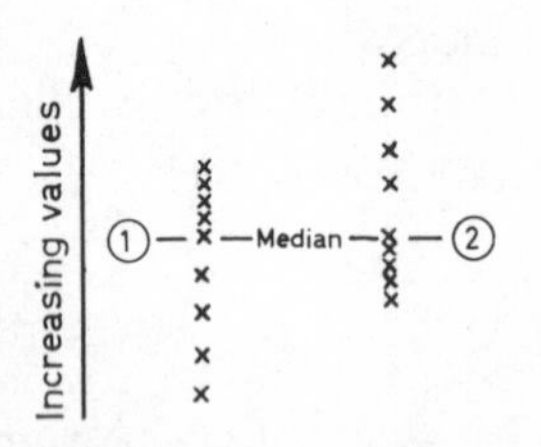

Figure 9. Relationship of the median to two sets of data

Of the three types of mean which have been considered, it is only the average which is based on sound mathematics, and which therefore possesses properties which permit its use in further calculations. Nevertheless, it is essential that the implications and limitations of the average also be appreciated. In the calculation of the average, weight is given to each occurrence according to its magnitude, in that all occurrences and their order of magnitude are used in its computation. Thus the extreme values are excessively stressed in comparison with the middling values. In a distribution which approximates to the normal (Fig. 3) this is of minor importance at the most, and in each of the three idealized distributions considered earlier (pp. 8–10)—normal, slight positive skewness and slight negative skewness—half of the occurrences exceeded the average and half were less than it. In cases of marked skewness, however, this will not be so. The following values may represent the annual falls of rain in inches in a desert area over ten years, and the resulting distribution curve is seen in Fig. 10.

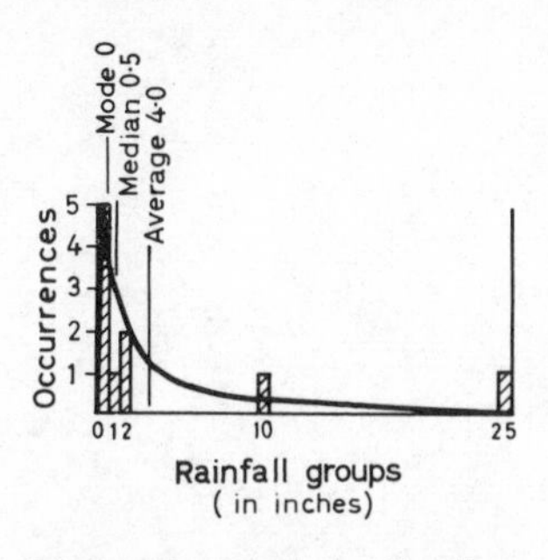

Figure 10. Frequency distribution curve and mean values for annual rainfall of a desert area

Fall (in.) = 0, 1, 0, 0, 10, 2, 25, 0, 0, 2;
total = 40 in.; average = 4·0 in.

Thus it can be seen that the average rainfall of 4″ was exceeded only twice in the ten years, while in the other eight years the rainfall was below the average. This is because the two wet years when the falls were 25″ and 10″ have each made a greater contribution to the total rainfall and have therefore affected the average value far more than have each of the more common years when rainfall was nil. In this particular example the median would be 0·5″ and the mode 0·0″, both of which give a better direct indication of the conditions which are most *typical*. Nevertheless, this does not imply that the average is useless or needlessly misleading in such cases, for any misinterpretation of the 4″ average value is a result of a failure by either the writer or the reader to appreciate what the average value really is and how it is calculated. On the other hand, this characteristic does illustrate one of the limitations of the average in relation to skew distributions, and also the possibly misleading character of *any* mean value when it is used alone. For a proper appreciation of the significance and relevance of any mean value it is also necessary to know something more of the distribution which the mean summarizes, i.e. it is desirable to know how actual conditions are 'scattered' around the mean value. It is the various methods by which this can be done, and their implications, to which attention must be paid in the next chapter.

CHAPTER 3

DEVIATION AND VARIABILITY

The fact that in any set of data the actual values differ from one another, and also from the mean value itself, has been stressed several times in the foregoing pages. A necessary corollary is that for a true and worthwhile understanding of the mean value of a set of data it must be possible to associate that mean easily and readily with some measure of the degree of scatter about that mean. If this can be achieved then the utility of the mean value is greatly increased, the effectiveness of the descriptive role of statistics is improved and many further deductions can be made concerning other properties of the set of data under consideration. Such applications in the field of geography will be presented in succeeding chapters.

Types of Deviation

If data are being presented graphically the simplest and most effective indication of scatter is provided by the frequency distribution curve, while a dispersion diagram in which each value is indicated is also useful. If scatter is to be expressed in numerical terms, however, these will not be applicable. One rough-and-ready way in which scatter can be expressed is in terms of the highest and lowest values occurring in the record. For example, the following two sets of figures both have the same average value, i.e. 5.

set i 1, 3, 5, 7, 9, average = 5
set ii 3, 4, 5, 6, 7, average = 5

Clearly the scatter about this average is different in the two cases and the ranges of values involved will give a very generalized idea of this. Thus it could be said that 'set i' has an average of 5 and a scatter from 1 to 9, while 'set ii' has an average of 5 and a scatter from 3 to 7. Although this is helpful in its own way, it is very imprecise. Moreover, it does not provide a summary of the scatter in *one* value only, which is desirable if statistical analysis is to be effective. It is therefore to methods which satisfy these conditions that attention must now be given.

Three more accurate and useful methods of summarizing scatter are commonly employed, these methods yielding numerical values which are referred to as 'deviation' values, i.e. values representing differences from the mean. Two of these methods can be used with the average, while one can be used with the median. If the mode is the type of mean value being employed, no effective deviation value can be presented, which is another disadvantage in the use of this value.

If conditions are being presented by the median then scatter can be summarized by the *quartile deviation.* This is derived just as simply as is the median itself, and it equally possesses the same advantages

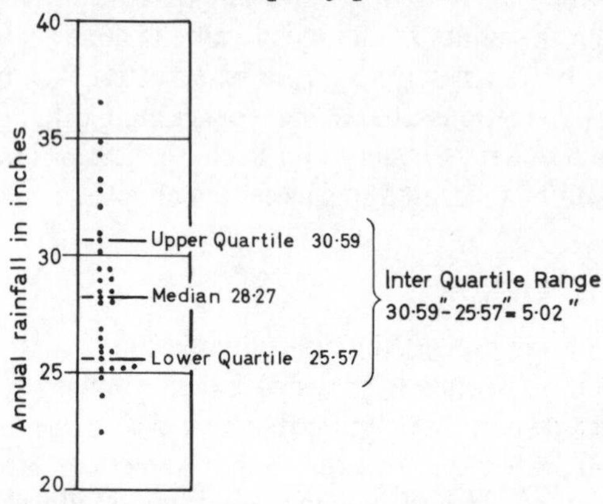

Figure 11. Graphical calculation of the median and quartiles for annual rainfall at Bidston, 1901–1930

and disadvantages as the median. The median was obtained (p. 7) by dividing the record into two equal parts as regards number of occurrences, this being effected by an inspection of either the figures themselves or a graphical plot of those figures. The two halves of the record, above and below the median, can then each be divided into two so that the overall record is divided into four groups of equal number of occurrences. The new dividing lines are called the upper and lower quartiles, the former separating the 25% of the record with the highest values from the rest, and the latter similarly separating the 25% with the lowest values. In Fig. 11 the values for rainfall at Bidston Observatory, which were set out in Table I, are presented graphically in order of magnitude. On this graph are entered the median at 28·27" as was obtained on p. 11, and also these two quartile values. Thus above the median there are 15 values of which the central one is the eighth from the top, so that the upper quartile is 30·59″. Again, there are 15 values below the median, the central one of these being the eighth from the bottom, i.e. the lower quartile is 25·57″.

These new quartile dividing lines enclose within them the central 50% of the occurrences. The difference between the top and the bottom of this central 50% is called the *inter-quartile range* which for the example in Fig. 11 is 30·59″ − 25·57″ = 5·02″. This range lies athwart the median, and if the distribution curve were normal and symmetrically balanced, then each of the quartiles would lie *half* this distance away from the median, i.e. 2·51″ away in the above example. It is this value, which gives an indication of the range of the central 50% of the occurrences above and below the median, that is called the quartile deviation. It may thus be expressed as

$$\frac{\text{upper quartile} - \text{lower quartile}}{2}$$

and can be described as the mean expectation of the deviation from the mean. In other words, half the occurrences differ from the mean (median) by *more* than this amount and half differ from it by *less* than this amount.

This is a very useful method, providing an easily-obtained value which possesses some clear meaning. On the other hand, it is still not really a true measure of overall scatter of the occurrences about the mean, for as in the case of the median the order of magnitude of the occurrences other than those specifically associated with the critical values, i.e. the quartiles, is not considered at all. It is only the existence of a given number of occurrences between or beyond certain points that is taken into account, not their order of magnitude. This characteristic once again renders the median/quartile system of only limited use for further computations. Nevertheless it is a valuable illustrative device which is widely used especially in the presentation of climatic data. The 10 mile to 1 inch rainfall map of the British Isles, prepared by the Meteorological Office for the Ministry of Town and Country Planning, provides an excellent example.

It was stressed in the previous chapter, however, that of the three means it is the arithmetic average which has most to commend it, and measurements of scatter in relation to it therefore need consideration. This can first be done in terms of a simple example. In the short set of data given below the average of the six values is 3·5:

$$6 + 5 + 4 + 3 + 2 + 1 = 21 \quad \text{Average } (\bar{x}) = \frac{21}{6} = 3{\cdot}5$$

A simple way of assessing the scatter of these values about this average is first to find out by how much each occurrence differs from (i.e. deviates from) this average value. These individual differences or deviations can be tabulated alongside the values themselves, as is done below. Once this has been set out it is a simple proposition to

Values (x)	Deviations (d)
6	2·5
5	1·5
4	0·5
3	0·5
2	1·5
1	2·5
6)21	6)9·0
$\bar{x}$ = 3·5	1·5

calculate the *average* amount by which individual values deviate from the mean. In other words, this gives the mean (average) deviation from the mean (average), and is known as the *Mean Deviation.* It is apparent, however, that no consideration has been given to the direction of these individual deviations, whether they be above or below the average. Instead, this question of direction or 'sign' (+ or −) has simply been ignored despite the fact that the sign is part of the mathematical quality of the deviations. This fact is recognized in the stricter definition of the mean deviation which can be presented as 'the average difference between various measurements and the central mean value, irrespective of sign'. This is written as follows:

$$\text{Mean deviation} = \frac{\Sigma \,|\, x - \bar{x} \,|}{n}$$

Here the fine vertical lines indicate that for this purpose it does not matter whether the value x is greater or less than the average $\bar{x}$, i.e. it is the difference between them *irrespective of sign* which is summed and averaged.

This ignoring of the sign is very convenient, making for simple calculation and easy understanding of the meaning of the resultant value. Nevertheless it is unsound, for the sign is necessarily an integral part of the value; moreover, the resultant deviation value is interpreted as *plus or minus* away from the average, as was the quartile deviation in relation to the median. If it is to be removed or standardized this should be effected by mathematical means. Therefore, before illustrating the mean deviation in terms of an actual set of data, it is desirable to outline the method by which the sign may be properly dealt with, so that the two methods can then be compared.

If the above example is reconsidered, a different way of removing the sign can be seen. This is by means of *squaring* all of the differ-

x	d	d^2
6	+2·5	6·25
5	+1·5	2·25
4	+0·5	0·25
3	−0·5	0·25
2	−1·5	2·25
1	−2·5	6·25
$\bar{x}$ = 3·5		6)17·50
		2·917
		$\sqrt{2{\cdot}917}$ = +/−1·7

ences, when the sign thus becomes positive in all cases. This is shown in the example under the column headed d^2. Then, as with the mean deviation, these several individual deviations are summed to give the *sum of the squares* and averaged. This value—the average of the squares of the deviations from the average—is known as the *variance* of the set of data, a characteristic to which reference will frequently be made in later sections. It represents the average amount of deviation from the mean, the negative signs having been changed to the *positive* by mathematical methods. As indicated above, however, a deviation value purports to summarize differences from the mean in both a positive *and* a negative direction. This feature can be introduced by finding the square root of this variance, for the square root of any number has both a positive and a negative value, i.e. $\sqrt{4} = +2$ or -2. So in the above example, while the variance is 2·917 the deviation value is + / − 1·7. Such a value is known as the *standard deviation*, or sometimes as the '*root mean square deviation*', the latter really explaining how it is calculated. Thus a full definition would be that 'the standard deviation is the square root of the average of the squares of the deviations from the arithmetic average'. This parameter is usually written as the Greek letter 'sigma'—σ, while the variance, the square of the standard deviation, is written as σ^2. The relationship between variance and standard deviation is fundamental to many later discussions and formulae, so it is essential to remember it. It is clearly apparent if the two formulae are written above one another:

$$\text{variance } (\sigma^2) = \frac{\Sigma (x - \bar{x})^2}{n}$$

$$\text{standard deviation } (\sigma) = \sqrt{\frac{\Sigma (x - \bar{x})^2}{n}}$$

The symbols used here are the same as have been used earlier, and careful working through the formulae in terms of the explanations given above will clarify what they mean.

Specific Examples

The application of these methods to some of the data which were considered in the previous chapter will also illustrate both the methods of calculation and some of the properties of the resulting deviation values. Their value in analysing geographical problems will then be outlined in succeeding chapters. Thus in Table III the rainfall data for Bidston Observatory are analysed to obtain both mean deviation and standard deviation. The methods used are those presented in the simple example above, and values of 2·79″ and 3·45″ are obtained for the mean and standard deviations respectively. The difference in magnitude between these is quite typical, the standard deviation being about 25% larger than the mean deviation. This approximate relationship, i.e. standard deviation = 1·25 mean deviation, is almost perfectly fulfilled in the case of the Bidston rainfall, for the factor by which the mean deviation must be multiplied to give the standard deviation proves to be 1·24. The relationship for annual rainfall in the British Isles, based on 230 stations, is shown in Fig. 12. Such a close similarity between actual and theoretical conditions does not always apply, of course, as the values in

Table III

The calculation of mean and standard deviation for annual rainfall at Bidston Observatory, Birkenhead, 1901–1930

Value	Deviation	Deviation squared
x	d	d^2
25·19	−3·26	10·6
25·57	−2·88	8·3
34·42	+5·97	35·6
25·18	−3·27	10·7
24·01	−4·44	19·6
28·08	−0·37	0·1
26·57	−1·88	3·5
28·90	+0·45	0·2
28·45	0·00	0·0
28·59	+0·14	0·02
25·27	−3·18	10·1
30·17	+1·72	2·95
25·78	−2·67	7·1
26·02	−2·43	5·9
26·83	−1·62	2·6
24·87	−3·58	12·8
30·59	+2·14	4·6
31·93	+3·48	12·1
29·12	+0·67	0·45
33·34	+4·89	23·9
22·47	−5·98	35·7
25·97	−2·48	6·15
30·92	+2·47	6·1
32·87	+4·42	19·6
28·00	−0·45	0·2
28·95	+0·50	0·25
34·81	+6·36	40·5
29·11	+0·66	0·4
25·15	−3·30	10·9
36·50	+8·05	65·0
30)853·63	30)83·71	30)355·92

Ave. = 28·45 Mean deviation = 2·79

Variance = 11·86 = σ^2

Standard deviation = $\sqrt{11{\cdot}86}$ = 3·45 = σ

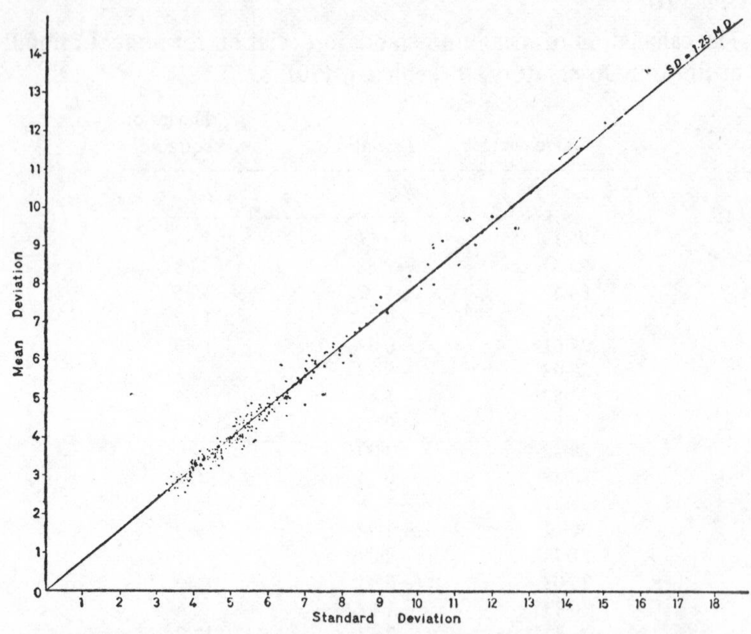

Figure 12. Graph of mean deviation against standard deviation values for annual rainfall for 230 stations in the British Isles

Table IV show. These values are for the iron-ore production figures which were earlier presented in Table II. It can be seen that the standard deviation is invariably greater than the mean deviation, but the proportions in these four cases vary between 1·16 and 1·37. Both

Table IV

Relationship between mean and standard deviations for iron-ore production in Belgium, France, Luxembourg and the United Kingdom, 1938–1957

Country	Average	Mean deviation	Standard deviation	SD/MD
	(thous. tons)	(thous. tons)	(thous. tons)	
Belgium	34·45	13·15	15·55	1·18
France	9,310·2	4,283·2	4,960·0	1·16
Luxembourg	1,448·8	437·3	527·2	1·20
United Kingdom	4,418·95	480·1	656·5	1·37

these features occur because during the process of squaring the individual deviations and then taking the square root of the *sum* of these squares, the larger deviations carry increased weight, while the smaller are given somewhat decreased weight. This means that the standard deviation will always be the greater of the two and that the degree of difference will be controlled by the relative frequency and magnitude of large and small individual deviations. Moreover, this also reflects the fact that these iron-ore data are all somewhat skew, while the Bidston rainfall data are almost normally distributed.

Alternative Formulae

It is the standard deviation which is the soundest indication of scatter in mathematical terms, and it is essential for other computations and formulae, as will be seen later. The considerable labour involved in its calculation if a calculator is not available is, however, something of a problem and a nuisance. Any short cut in the process of calculation is therefore to be welcomed, and one of these is based on an algebraic modification of the formula, so that the number of individual calculations is decreased. This not only saves time but also reduces the possibilities of error.

The formula for the variance has been shown to be the following (p. 24):

$$\sigma^2 = \frac{\Sigma (x - \bar{x})^2}{n}$$

The major component of this, and the portion that involves the bulk of the calculations, is $(x - \bar{x})^2$, and it is possible to write this out in full in the following manner:

$$(x - \bar{x})^2 = (x - \bar{x})(x - \bar{x}) = x^2 - 2\bar{x}x + \bar{x}^2$$

This therefore allows the formula for the variance to be rewritten, i.e.

$$\sigma^2 = \frac{\Sigma x^2}{n} - \frac{\Sigma 2\bar{x}}{n} \cdot \frac{\Sigma x}{n} + \frac{\Sigma \bar{x}^2}{n}$$

Thus each separate part of the expanded version of σ^2 can be summed and then divided by the number of occurrences. Although this is apparently a more complicated and confusing version, it is nevertheless possible to simplify the individual components. On

p. 6 it was shown that the summation of x over n gives the average, i.e.

$$\frac{\Sigma x}{n} = \bar{x}$$

and therefore $\bar{x}$ can be substituted for $\frac{\Sigma x}{n}$. Again, as $\bar{x}$ is the average it is bound to be a constant, i.e. always the same, in the one formula. Therefore if $\bar{x}$ is added up n times and then divided by n, the answer must be $\bar{x}$, i.e.

$$\frac{\Sigma \bar{x}}{n} = \bar{x}$$

It is now time to attempt the simplification of the formula for the variance in the following way, by the substitution of $\bar{x}$ for both $\frac{\Sigma x}{n}$ and $\frac{\Sigma \bar{x}}{n}$

$$\sigma^2 = \frac{\Sigma x^2}{n} - \frac{\Sigma 2\bar{x}}{n} \cdot \frac{\Sigma x}{n} + \frac{\Sigma \bar{x}^2}{n}$$

$$= \frac{\Sigma x^2}{n} - 2\bar{x}.\bar{x} + \bar{x}^2$$

$$= \frac{\Sigma x^2}{n} - 2\bar{x}^2 + \bar{x}^2$$

$$= \frac{\Sigma x^2}{n} - \bar{x}^2$$

Furthermore, as the standard deviation is simply the square root of the variance, then the standard deviation formula can be written

$$\sigma = \sqrt{\frac{\Sigma x^2}{n} - \bar{x}^2}$$

This involves far fewer calculations. Each individual occurrence is squared, these values are summed and averaged, and then the square of the average of the occurrences is subtracted. Finally, the square root of this must be obtained to change it from the variance to the standard deviation.

A practical example will make a clearer distinction between the two methods. Suppose that a study is being made of the sphere of influence of a particular town. Amongst the aspects of this that might be studied could well be the frequency of train services to neighbouring centres of population. From such a study assume that it was found that 25 such centres were served, and the number of trains per day to each of these centres were as set out in the second column of Table V. By simple calculation it could be found that the average number of trains per day between the town being studied and any neighbouring centre was 9·6. Apart from the need for careful use of such a figure because the set of data consists of a discrete rather than a continuous variate, it would also be useful to know the scatter of the values about this mean, preferably in terms of the standard deviation.

On the right-hand side of Table V the variance is calculated by the first of the formulae to be presented above, i.e. by $\sigma^2 = \frac{\Sigma (x - \bar{x})^2}{n}$ (METHOD 1), while on the left-hand side the second of the formulae is used, i.e. $\sigma^2 = \frac{\Sigma x^2}{n} - \bar{x}^2$ (METHOD 2). As can be seen, they both give the same variance value, i.e. 26·0, so that the standard deviation in each case is 5·1. The number of calculations involved is markedly different, however. In Method 1 there are 25 subtractions, 25 squares, 1 addition, 1 division and 1 square root—a total of 53 operations, each of them a source of possible error and a consumer of time. In Method 2 the total number of calculations is reduced to 30, i.e. 26 squares, 1 addition, 1 division, 1 subtraction and 1 square root. Moreover, until the final phases the values involved are all whole numbers without decimals. This is only true, however, because the problem is concerned with a discrete rather than a continuous variate, although this will not always be the case. On the other hand, the size of the numbers involved in Method 2 can prove to be very large indeed. The method is most valuable, in fact, when some form of calculating machine is available, for even with a small hand-operated desk adding machine it is possible to calculate both the average *and* the standard deviation *at the same time*. There is no equivalent short cut in the mechanical handling of Method 1. By both methods the variance is shown to be 26·0 and therefore the standard deviation is the square root of this

i.e. $\sigma = \underline{\underline{5{\cdot}1}}$

Table V

The calculation of the standard deviation by two methods using data concerning the number of trains per day between one town and neighbouring towns

(METHOD 2—p. 28) Occurrences squared	No. of trains per day	(METHOD 1—p. 24) Difference	Difference squared
x^2	x	$(x - \bar{x})$	$(x - \bar{x})^2$
1	1	− 8·6	73·96
4	2	− 7·6	57·76
9	3	− 6·6	43·56
9	3	− 6·6	43·56
16	4	− 5·6	31·36
25	5	− 4·6	21·16
36	6	− 3·6	12·96
36	6	− 3·6	12·96
64	8	− 1·6	2·56
64	8	− 1·6	2·56
64	8	− 1·6	2·56
100	10	+ 0·4	0·16
100	10	+ 0·4	0·16
100	10	+ 0·4	0·16
100	10	+ 0·4	0·16
121	11	+ 1·4	1·96
121	11	+ 1·4	1·96
144	12	+ 2·4	5·76
144	12	+ 2·4	5·76
196	14	+ 4·4	19·36
225	15	+ 5·4	29·16
225	15	+ 5·4	29·16
289	17	+ 7·4	54·76
361	19	+ 9·4	88·36
400	20	+ 10·4	108·16
25)2,954	25)240		25)650·00

$$\frac{\Sigma x^2}{n} = 118{\cdot}16 \qquad \bar{x} = 9{\cdot}6 \qquad \sigma^2 = \frac{\Sigma (x - \bar{x})^2}{n} = \underline{\underline{26{\cdot}0}}$$

$$\bar{x}^2 = 92{\cdot}16 \qquad \sigma^2 = \frac{\Sigma x^2}{n} - \bar{x}^2 = 118{\cdot}16 - 92{\cdot}16 = \underline{\underline{26{\cdot}0}}$$

If a calculating machine is being used, however, perhaps the best modification of the formula to use is the following:

$$\sigma^2 = \frac{1}{n} \cdot \left(\Sigma x^2 - \frac{(\Sigma x)^2}{n}\right)$$

As in method 2, the initial calculation involves simply summing the original values ($\Sigma x = 240$) and summing these values squared ($\Sigma x^2 = 2{,}954$), which can normally be done in one operation. The calculation then becomes

$$\sigma^2 = \frac{1}{25}\left(2954 - \frac{(240)^2}{25}\right) = \frac{1}{25} \times 650$$
$$= 26{\cdot}0$$

In this third method there is thus no need to calculate the average at all (although the basic values of Σx and n are there if needed), while the value for the 'sum of the squares' is separately obtained in case it is needed later.

Grouped Frequency Method

If even the simpler mechanical aids are not available to speed up the work, then a long series of data may be analysed by a more generalized method which gives an answer approximating very closely to the correct one. Moreover, this method also allows for the calculation of the average in the same generalized way and *at the same time*. The method involves certain 'rounding' or simplifying processes which may at first seem arbitrary and unjustified, but the proof of the general accuracy of the method can be readily demonstrated by a practical example—in fact, by reworking the data which have just been analysed by the three exact methods. Mathematical proofs of the adequacy of this method are also possible, but will not be presented here—the important thing is to become familiar with the technique simply as a useful tool.

The tabulation, formulae and calculations necessary are set out in Table VI which should be followed carefully in connection with the explanation which follows. The first task is to group the data into classes or cells, as is done in the preparation of a histogram (p. 2). It is essential, however, that for this purpose the range of values in any one class should be small and that the number of classes should be at least 10. If these two conditions are not satisfied the margin of error introduced by the generalization may well be too large to allow the answers to be of any real use. In the present example, where the variate is discrete and all the data must be in the form of whole numbers, it is in many ways adequate simply to list the 'class marks' which are shown at the beginning of Table VI. The classes here are 10 in number, each of them consisting of two numbers. If the example were a continuous variate then the classes would need to cover all contingencies, and class boundaries would need to be carefully defined. Even in a case such as the present it is desirable to establish class boundaries, as this facilitates the correct interpolation of the class mid-marks. A certain amount of care and thought is required when these class boundaries are being defined. Thus in Table VI it could be argued that all values between 0·5 and 1·0 are rounded to 1, and that all values between 2·0 and 2·5 (but not including 2·5 itself) are rounded to 2. Therefore the boundaries of this class are from 0·5 to 2·5 and those for this and all the other classes are shown in the second column in Table VI. The correct definition of the class mid-mark is now easier, as it is the central value within the class boundaries. In the present example these mid-marks are thus 1·5, 3·5, 5·5 etc. up to 19·5. These now become the values of the occurrences with which these calculations are made, and they are entered in the third column under the heading x to indicate this. With these entered, it is easy to see the magnitude of the difference between mid-marks, this being known as the 'cell or class interval' and written as c—here it is 2. Finally in this preparatory tabulation it is necessary to enter in the fourth column, under the heading f, the number of occurrences falling within the boundaries of each class, i.e. the frequency of occurrence needs to be obtained, the total frequency Σf being entered at the bottom of the column.

It is with the cells, mid-marks and frequencies which are thus established by simple inspection of the data that this computation of average and standard deviation values is concerned. The actual values themselves, and the possible errors of calculation which result from working with complex numbers, are now temporarily discarded and these small simple numbers are used instead. To do this it is necessary first to adopt an *assumed* mean value, choosing, if possible, the mid-mark closest to the actual arithmetic mean. This is largely a matter of experience, but it does not matter in any fundamental sense if the mid-mark chosen as the assumed mean is in fact markedly different from the actual mean. This will not invalidate the answer obtained, nor will it necessitate any change in method of computation. Its sole effect is that the resultant calculations will involve larger numbers than would otherwise be necessary. In the present example the assumed mean (indicated by $\bar{x}_0$) is entered as 11·5.

Table VI

The calculation of the average and the standard deviation by the grouped frequency method, using the same train per day data as in Table V

Class marks	Class bounds	Class mid-mark	Frequency	Deviation of cell from $\bar{x}_0$ in units of c	Total deviation of class	Total of deviation squared
		x	f	t	ft	ft^2
1– 2	0·5– 2·5	1·5	2	−5	−10	50
3– 4	2·5– 4·5	3·5	3	−4	−12	48
5– 6	4·5– 6·5	5·5	3	−3	− 9	27
7– 8	6·5– 8·5	7·5	3	−2	− 6	12
9–10	8·5–10·5	9·5	4	−1	− 4	4
11–12	10·5–12·5	11·5	4	0	0	0
13–14	12·5–14·5	13·5	1	+1	1	1
15–16	14·5–16·5	15·5	2	+2	4	8
17–18	16·5–18·5	17·5	1	+3	3	9
19–20	18·5–20·5	19·5	2	+4	8	32
Assumed mean $\bar{x}_0$ = 11·5			25		−25	191
Cell interval c = 2			Σf		Σft	Σft^2

Arithmetic average:

$$\bar{x} = \bar{x}_0 + c.\frac{\Sigma ft}{\Sigma f}$$

$$= 11{\cdot}5 + 2\left(\frac{-25}{25}\right) = 11{\cdot}5 + (-2)$$

$$= 11{\cdot}5 - 2 = \underline{\underline{9{\cdot}5}}$$

Standard deviation:

$$\sigma = c.\sqrt{\frac{\Sigma ft^2}{\Sigma f} - \left(\frac{\Sigma ft}{\Sigma f}\right)^2}$$

$$= 2.\sqrt{\frac{191}{25} - \left(\frac{-25}{25}\right)^2} = 2.\sqrt{7{\cdot}64 - 1}$$

$$= 2\sqrt{6{\cdot}64} = 2 \times 2{\cdot}58$$

$$= \underline{\underline{5{\cdot}16}}$$

It is now time to begin the real calculation, having transferred the numerical data into a suitable form. Deviation from the mean is calculated not in absolute terms but rather as the number of 'units of cell interval' away from the *assumed* mean, i.e. the number of c units away from the class mid-mark chosen as $\bar{x}_0$, and these are entered in the fifth column under t. The class with the mid-mark equal to the assumed mean has a value of 0 entered under t, indicating that the class as a whole is being considered as equal to the mean. Other values range successively as negative values (−1, −2, −3 etc.) and positive values (+1, +2, +3 etc.) in the appropriate directions. This gives the amount by which the *cell* deviates from the assumed mean in terms of units of cell intervals (c). To obtain the total deviation within any given cell it is necessary to multiply this deviation by the number of occurrences within that cell, i.e. to multiply t by f. This value of ft is entered in the sixth column in Table VI, the total deviation of the whole series being entered at the foot of the column as $\sum ft$.

The calculation of the average from these retabulated data is now possible. It will normally be found that the assumed mean differs to some extent from the actual mean, although the amount of this difference is not known in advance. The correction for this difference is simply obtained, however. In column six of the tabulation is given the total amount by which the actual data differ from the assumed mean. If this value Σft is divided by the total number of occurrences, i.e. by Σf, the amount by which the actual average differs from the assumed average is obtained. This value is given here in units of cell intervals and must therefore be multiplied by this cell interval value, i.e. by c, to transfer this difference into actual numbers. By adding this difference to the assumed mean the actual mean is obtained. The formula for this calculation is set out in Table VI. In this example it is found that the assumed mean differs from the actual mean by one cell interval, i.e. $\frac{\Sigma ft}{\Sigma f} = -1$. As the cell interval is a value of 2 then the assumed mean must be adjusted by −2 to give the actual mean. The calculations indicate that this adjustment must be by subtraction, for the assumed value is *higher* than the actual one, so that the actual mean is given as 9·5. This is a very close approximation to the true value obtained by normal calculations (Table V), which was 9·6. This value of 9·5 obtained by the *grouped frequency method*, as it is called, is exactly the same as one of the class mid-marks. If this value had been chosen as the assumed mean, which could quite easily have been the case, the total deviation value under Σft would have been 0—a simple calculation will show this to be true. In that case the adjustment to be applied to the assumed mean, under the factor $c.\frac{\Sigma ft}{\Sigma f}$, would also have been 0, thus giving the same answer as obtained in Table VI. This also stresses the point made earlier that the closer the assumed mean is to the actual mean, the easier the calculations that need to be made. As for the difference between the mean value by this method and that which is correctly obtained by the normal method of calculation, this results mainly from the fact that only 10 cells were used, this being the minimum desirable number. If the number of these had been larger, and the size of the cells thus smaller, then the difference between the two methods would have been less. As it is, the difference is no greater than one decimal place, and an accuracy as great as this with as little involved calculation will prove of inestimable value in the case of sets of data comprising several hundreds of occurrences.

This account of the grouped frequency method of calculating the average has been something of a digression, but as in practice the average is usually calculated from the same tabulation as is the standard deviation, its inclusion at this point is pertinent. To obtain the standard deviation by this method requires some further calculation. As with the ordinary method of calculation, the sum of the *squares* of the deviations is needed, i.e. the deviation t is squared and then multiplied by the frequency in the cell f. This is obtained from the seventh column in Table VI, where this value is given in terms of cells, under the head Σft^2. Again applying the standard formula this value has to be averaged, i.e. divided by Σf, to give the variance, from which the standard deviation can be obtained by taking the square root. In the present method, however, these deviations have been measured from an *assumed* mean, so that as in the case of the average outlined above a correction must be applied for this. This correction is the same as for the average, i.e. $\frac{\Sigma ft}{\Sigma f}$, save that this value must be squared to ensure that it is of the same proportions as the deviations which were themselves squared. Once this is subtracted from the mean of the sum of the squares, the square root can be obtained, thus giving the standard deviation in cells. To obtain the correct answer this value must now be multiplied by the cell interval. This descriptive account is most clearly appreciated if it is closely followed in Table VI, bearing in mind that the formula given is fundamentally the same as Method 1 presented for the standard deviation on p. 24. The only differences are that a correction factor is introduced to allow for the assumed mean and the answer has to be multiplied by the cell interval because all calculations are in terms of cells. In the example in Table VI the standard deviation is given as 5·16, while the answer by the usual method of calculation is 5·10. Again, as with the average, the margin of error is very small, despite the limited number of cells used. The amount of time saved, if the number of occurrences is considerable and if they include large and awkward values, is such that the small error is usually worth accepting.

As this method may appear rather complicated, although in reality it proves very simple to work, a clearer understanding may be obtained if another example is presented, this time with greater numbers involved. The problem to be analysed can be outlined in the following way. During a study of farming it is found that poultry plays a part in the economy of all the sample farms in the area. The number of poultry kept varies considerably, however, from as low as 2 to as high as 200, and it is desired to discover the average number of poultry per farm and also the standard deviation of this set of values. The data are set out in tabular form in Table VII. Twenty cells are defined, from 1–10 to 191–200, the cell interval, i.e. c, being 10. The mid-marks of each cell are also defined, these being 5·5, 15·5, 25·5 etc. up to 195·5, and they are entered under x. In the following column is shown, under f, the frequency with which occurrences fall within the given cells, and it is seen that there are 1,044 occurrences altogether, i.e. $\Sigma f = 1{,}044$. With a number such as this the full calculation of average and standard deviation values would obviously be a lengthy process. As the frequency distribution appears to be a relatively normal one, the assumed mean is chosen at about the central point so as to keep the size of numbers to a minimum. It is therefore taken as 105·5. The deviation of each cell from this value, in terms of the number of cell intervals, is then entered under t, the value 0 being entered against the cell with the same value as the assumed mean, while values of -1, -2 etc. extend to the cells with progressively lower mid-marks and values of

Table VII

The calculation of the average and the standard deviation by the grouped frequency method, and the application of Charlier's Test, using data about the number of poultry per farm

Class Marks	Class Mid-marks	Fre-quency	Deviation of Cell	Total deviation	Total deviation squared	For test
	x	f	t	ft	ft^2	$f(t+1)^2$
1– 10	5·5	5	−10	− 50	500	405
11– 20	15·5	12	− 9	−108	972	768
21– 30	25·5	19	− 8	−152	1,216	931
31– 40	35·5	24	− 7	−168	1,176	864
41– 50	45·5	33	− 6	−198	1,188	825
51– 60	55·5	52	− 5	−260	1,300	832
61– 70	65·5	69	− 4	−276	1,104	621
71– 80	75·5	75	− 3	−225	675	300
81– 90	85·5	108	− 2	−216	432	108
91–100	95·5	120	− 1	−120	120	0
101–110	105·5	123	0	0	0	123
111–120	115·5	101	+ 1	101	101	404
121–130	125·5	85	+ 2	170	340	765
131–140	135·5	79	+ 3	237	711	1,264
141–150	145·5	60	+ 4	240	960	1,500
151–160	155·5	43	+ 5	215	1,075	1,548
161–170	165·5	21	+ 6	126	756	1,029
171–180	175·5	9	+ 7	63	441	576
181–190	185·5	4	+ 8	32	256	324
191–200	195·5	2	+ 9	18	162	200
Assumed mean $\bar{x}_0 = 105{\cdot}5$		1,044		−571	13,485	13,387
Cell interval $c = 10$		Σf		Σft	Σft^2	$\Sigma f(t+1)^2$

Charlier's Test:

$$\Sigma f(t+1)^2 = \Sigma ft^2 + 2\Sigma ft + \Sigma f$$

i.e. $13{,}387 = 13{,}485 - 1{,}142 + 1{,}044$

Arithmetic Average:

$$\bar{x} = \bar{x}_0 + c.\frac{\Sigma ft}{\Sigma f}$$

$$= 105{\cdot}5 + 10.\frac{-571}{1{,}044} = 105{\cdot}5 - 5{\cdot}46 = \underline{\underline{100{\cdot}04}}$$

Standard deviation:

$$\sigma = c\sqrt{\frac{\Sigma ft^2}{\Sigma f} - \left(\frac{\Sigma ft}{\Sigma f}\right)^2}$$

$$= 10\sqrt{\frac{13{,}485}{1{,}044} - \left(\frac{-571}{1{,}044}\right)^2} = 10\sqrt{12{\cdot}9 - 0{\cdot}299} = 10\sqrt{12{\cdot}6}$$

$$= 10 \times 3{\cdot}55 = \underline{\underline{35{\cdot}5}}$$

+1, +2 etc. to those with progressively higher mid-marks. The total amount of deviation in each cell is then given under ft and the total deviation from the assumed mean in the whole record is given as Σft. Also, in preparation for the standard deviation calculation these deviations are squared for each cell and this value again multiplied by the frequency in that cell, the answer being shown under ft^2 for each cell and under Σft^2 for the full record.

From this point the calculation is both simple and speedy. The average $\bar{x}$ is obtained by adding a correction to the assumed mean $\bar{x}_0$. This correction is the average amount by which each occurrence differs from the assumed mean, this being zero if the true and the assumed means are the same. As this correction is in terms of cells it must be multiplied by the cell interval c before being added to the assumed mean, which is in actual values. For the present example the calculation of the mean by this method is set out below Table VII, and an answer of 100·04 is obtained. As for the standard deviation, the mean of the sum of the squares of the deviations from the *assumed* mean, expressed in cells, is given by $\frac{\Sigma ft^2}{\Sigma f}$; this is corrected to deviations from the actual mean by subtracting the above correction factor squared; and the standard deviation is obtained when the square root is calculated for this amount and converted from cells into actual values by multiplying by the cell interval. The answer in this case is seen to be 35·5.

When a calculation of this sort is being made, however, there is always the possibility of arithmetical errors creeping in. It is therefore desirable to institute a check upon the accuracy of the calculations in the tabulation. In the present case this is most easily provided by the application of what is known as *Charlier's Test*. As the result of some slight increase in calculation this test indicates whether the main body of the working has been carried out properly. It must be admitted that it is not an infallible test, as it is possible for some compensation of errors to occur within the test, but this is extremely unlikely. As can be seen in Table VII, this test is applied by

$$\Sigma f(t+1)^2 = \Sigma ft^2 + 2\Sigma ft + \Sigma f$$

To obtain the value $\Sigma f(t+1)^2$, the simplest method is to add an extra column to the table. For each cell one digit is added to the t value; this is squared and then multiplied by the f value. Ideally, this calculation should be carried out before the average and standard deviation values are worked out, so that any corrections can then be made and needless labour avoided. This has been done in Table VII, where both sides of the equation are seen to be the same (13,387 in this case) thus indicating that the numerical calculations in the tabulation have been carried through accurately. Any small errors that remain are simply the result of the generalizing process on which this method is based.

Variability Indices

In all these assessments of deviation which have so far been considered, the deviation value has been expressed in absolute terms—that is to say, in terms of so many inches of rainfall, so many tons of iron-ore, so many trains per day per town or so many poultry per farm. Within any body of data, however, the magnitude of this value is at least in part controlled by the magnitude of the mean value. This can be seen in Table IV which has already been considered, and also in Fig. 13*a*, where the mean deviation of annual rainfall for some 230 stations in the British Isles is plotted against the mean values for those stations. It is when comparisons are being made that the influence of the magnitude of the mean value may be rather inconvenient. On other occasions, too, it is useful to be able to consider the relationship between deviation and the mean value itself.

In the case of all three of the deviation values which have been outlined above (standard, mean and quartile deviations) it is possible to indicate this relationship in the same way. This is by simply expressing the deviation value as a percentage of the requisite mean, thus eliminating the direct influence of the magnitude of the mean and facilitating comparison in relative terms between various sets of data. This resultant value can be regarded in two ways, both of

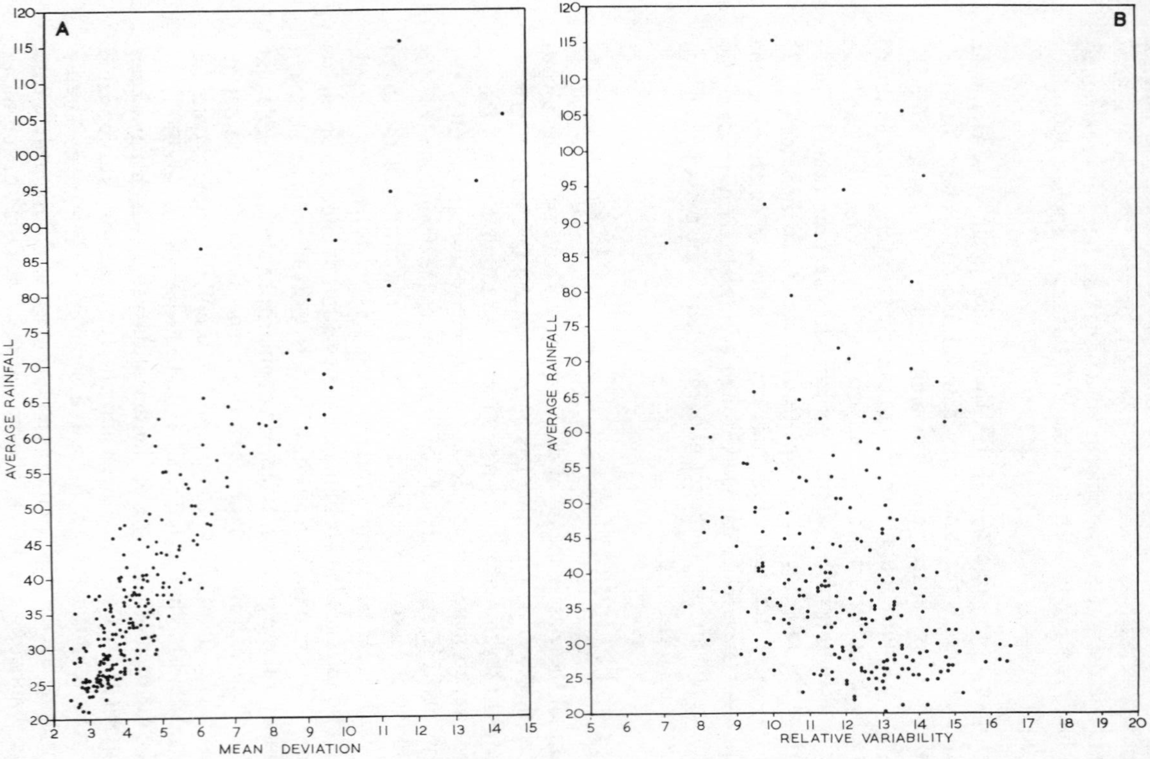

Figure 13. Graphs of mean deviation and relative variability values against the average for annual rainfall for 230 stations in the British Isles

which are valid. On the one hand, it represents the percentage variability of the set of data obtained in the way suggested above. On the other hand, if each individual deviation is expressed not in absolute terms but as a *percentage* deviation from the mean, then the whole calculation of the standard deviation, for example, could be made in these terms. In this way would be obtained the *percentage* standard deviation.

It is as an *index of variability*, however, that this percentage value is most often used by geographers, especially when distribution maps of variability are required. The elimination of the direct influence of the mean is its great value in these cases, as can be seen in Fig. 13*b*. Here the rainfall data from Fig. 13*a* are presented in terms of percentage rather than absolute values, and the removal of any direct relationship with mean value is clearly to be seen.

If the median and quartile deviation values are being used, an index of variability can thus be obtained easily by

$$\frac{\text{quartile deviation}}{\text{median}} \times 100\%$$

In terms of the Bidston rainfall data set out in Fig. 11 and Table I, this calculation becomes

$$\frac{2{\cdot}51}{28{\cdot}27} \times 100\% = 8{\cdot}9\%$$

Equally, the resultant values for the iron-ore production listed in Table II are given in Table VIII, along with the similar values for other methods to facilitate comparison.

Table VIII

Indices of variability for the iron-ore data previously presented in Table II

Country	Quartile deviation / Median	(Relative Variability) Mean deviation / Average	(Coefficient of Variation) Standard deviation / Average
	%	%	%
Belgium	41·3	38·2	45·1
France	44·85	46·0	53·3
Luxembourg	28·1	30·2	36·4
United Kingdom	6·65	10·8	14·85

When the average is being used, the index of variability depends on the deviation value. If the simpler mean deviation is employed, the variability value

$$\frac{\text{mean deviation}}{\text{average}} \times 100\%$$

is usually referred to as the *relative variability*. On the other hand, if it is the standard deviation which is used, so that the calculation is

$$\frac{\text{standard deviation}}{\text{average}} \times 100\%$$

then this value is known as the *coefficient of variation* and is written in formulae as V.

These three methods possess the advantages and disadvantages which are implicit in the values on which they are based, and which have been outlined earlier (pp. 14–24). In brief, this means that it is the coefficient of variation which is mathematically most correct and which therefore has the greatest *potential* value for assessing yet further the characteristics of the data under review. It is the relative variability which has been most widely used by geographers, however, its easier and quicker calculation in the absence of mechanical aids being a great advantage provided that no further calculations are intended. Moreover, the fairly close relationship between mean and standard deviations indicated in Fig. 12 means that isopleth maps based on these two different methods of indicating variability usually present virtually the same pattern—although the *quantitative* picture is necessarily different. Thus if only a relative comparison between areas is required the simpler method may be preferable, but if the results are to be used to assess, for example, the probability of certain conditions obtaining (Chapter 5) then the coefficient of variation is essential.

The different answers which are obtained by these three indices of variability are shown in Table VIII, where the iron-ore data used previously are employed. Several relevant points are stressed by this table. Firstly, it can be seen that the order of the countries in terms of the magnitude of variability is the same whichever method is used —France, Belgium, Luxembourg and the United Kingdom. Secondly, it is equally clear that the values differ markedly between the methods. This means that whenever variability is being presented it is essential that the method by which it is assessed be clearly stated. In most cases the method using the quartile deviation gives the lowest value, though this is not invariably the case, e.g. Belgium in this table. The failure to maintain a regular place in the order is the result of the limitations associated with the median and quartile deviation indicated on p. 21. With the other two methods, however, the coefficient of variation is always greater than the relative variability value, the difference between them being of the order of 25% as in the case of the mean and standard deviation (p. 24). The third major point from Table VIII is that in comparison with Table IV, where the deviation values are shown, the relative position of these four countries has changed, e.g. although the standard deviation for the U.K. is the second highest, its coefficient of variation is the lowest, i.e. the influence of the magnitude of the average has been removed.

In all these calculations, however, it must be remembered that a normal frequency distribution is assumed as is done for the majority of statistical methods (p. 8). This does not always occur, and therefore an index of overall variability will not adequately reflect the different tendencies and degrees of variability above and below the mean. This is again of major importance if further calculations of probability are to be made (Chapter 5).

As will become apparent later in Chapter 4 (p. 52), once the coefficient of variation exceeds 35% it becomes increasingly misleading, for under such conditions the data do not fit the normal frequency curve. Thus, as an extreme example, the case of January rainfall at Zaria, northern Nigeria, may be quoted. The coefficient of variation for a fifty-six year period is approximately 500%—apparently rainfall in January is extremely variable. Yet, in fifty-two of the years no rain at all fell in January, and in none of the four other Januarys did it exceed 0·04 in.—a lesser degree of variability can scarcely be visualized.

In several ways, therefore, the scatter of actual values about the mean can be calculated and expressed. These calculations are of varying degrees of complexity and can be made to various degrees of accuracy, depending on the method used. The chief decision that any individual has to make, however, is in what way the scatter of values should be expressed for any particular set of data. Thus the variability of annual rainfall over the British Isles can be expressed either by the Relative Variability (Fig. 14) or by the Coefficient of

Variation (Fig. 15). The issues involved include the purpose for which the calculation is being made, whether any further calculations are to be based on the results, the nature of the original data, the audience

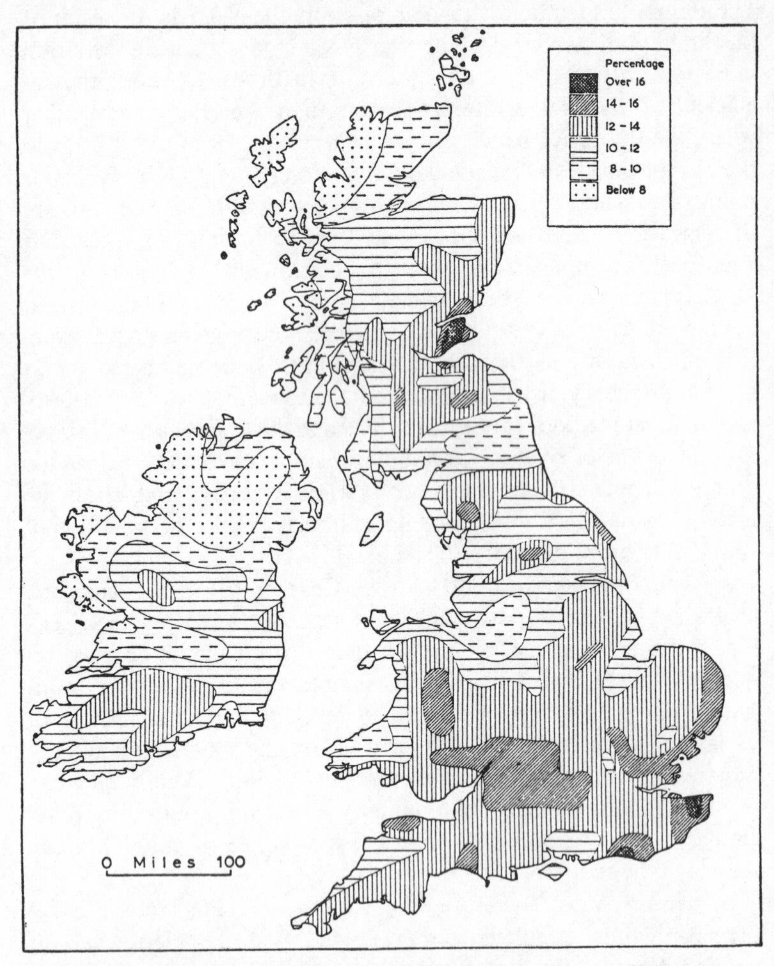

Figure 14. The relative variability of annual rainfall over the British Isles, 1901–1930 (from S. Gregory, *Quart. J. R. Met. Soc.*, 81 (1955))

to whom the results are to be presented, the presence or absence of mechanical or other aids to computation, and the degree of accuracy required. Decisions on these and other matters will control which of the methods presented in this chapter should be used in any particular case. It must be appreciated, however, that both the quartile and the mean deviation, as well as their associated indices of vari-

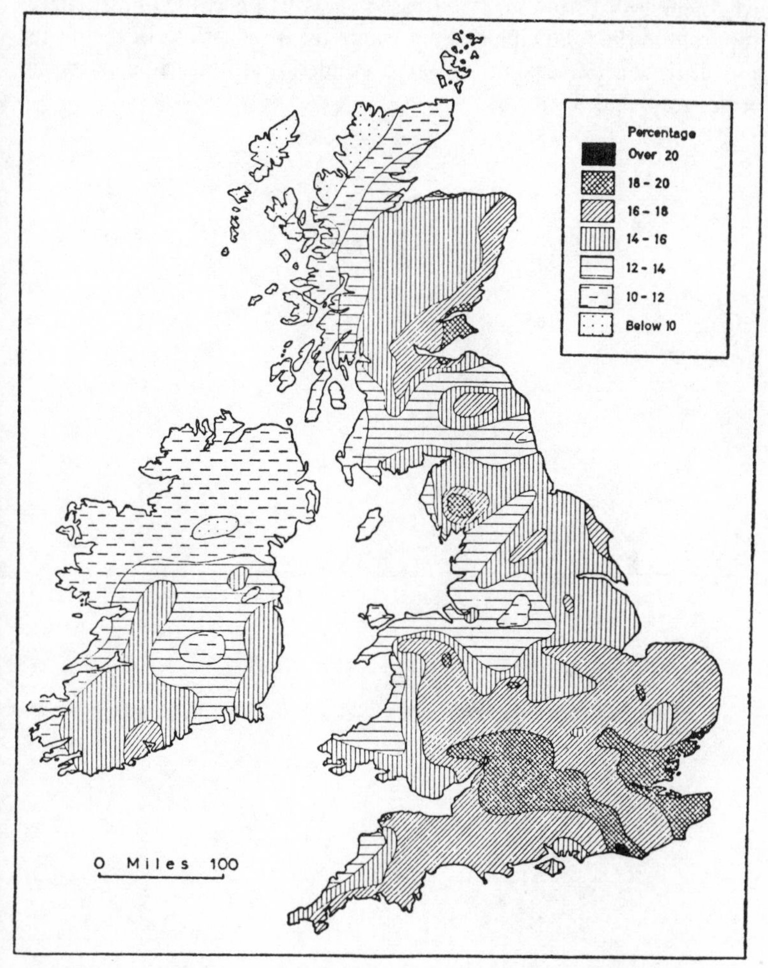

Figure 15. The coefficient of variation of annual rainfall over the British Isles, 1901–1930

ability, do not lend themselves to the assessment of further characteristics of the data, and that they (like the median and the mode as mean values) have thus only limited use. In the presentation of further methods of analysis in the rest of this book it is therefore upon the arithmetic average, the variance, the standard deviation and the coefficient of variation that both calculations and theoretical arguments must necessarily be based. This will become immediately apparent in the following chapter where the implications of the mean and deviation parameters will be presented and illustrated in terms of geographical problems.

CHAPTER 4

THE NORMAL FREQUENCY DISTRIBUTION CURVE AND ITS CHARACTERISTICS

In the previous chapters it has been shown that in order to represent a body of data adequately *two* parameters or characteristics need to be defined—a measure of central tendency and a measure of scatter or deviation—and moreover, it has been argued that of the various ways by which this might be done, the most effective and soundest method is by the use of the arithmetic average and the standard deviation. In these two values the body of data is briefly but satisfactorily summarized. There are also two other characteristics that are relevant—skewness and kurtosis—and these will be examined briefly at the end of this chapter.

Characteristics of the Normal Curve

If it were stated that the average yield of wheat per acre for a series of farms was 30 bushels and that the standard deviation was 5 bushels, what would this imply? What extra information could be interpreted from such a statement? The point that must be borne in mind is that the standard deviation presents a summary of the distribution curve of the data concerned, while the mean indicates the actual value about which this curve is distributed. On the assumption that the frequency distribution is a normal one (which is the usual assumption in statistical methods unless otherwise specified), this curve which the standard deviation represents is symmetrically placed about the central point which the average indicates. Thus if in several records the means differ but the standard deviations are the same, then the shape of the distribution curve will be the same in all cases but related to a different point on the magnitude scale—this is portrayed diagrammatically in Fig. 16. Conversely, if the average is kept constant while standard deviation values differ, different curves are indicated around the same central value. Again, these differences are seen in Fig. 16.

Within the area enclosed by each of these curves and the base line are recorded all the occurrences which contribute to the mean value, these being accounted for in terms of both their order of magnitude and the number of occurrences at each such order of magnitude. If, as stated above, the standard deviation summarizes the shape of the distribution curve then it equally summarizes the number of occurrences at each order of magnitude. The point is that given a normal distribution curve it is possible to postulate the number of occurrences at any given value and between given values. The mathematics of this are best left on one side—it is, in fact, described by the function

$$y = \frac{1}{\sqrt{2\pi}} . e^{\frac{-x^2}{2}}$$

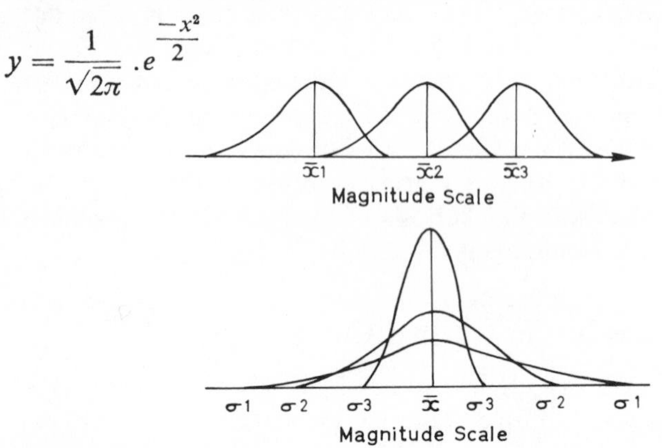

Figure 16. The graphical representation of changes in average and standard deviation values

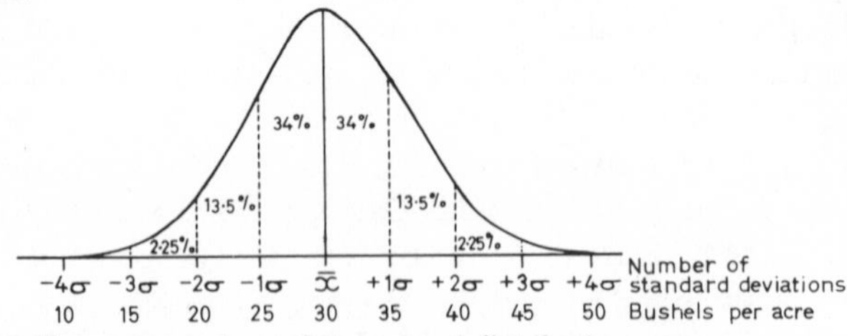

Figure 17. Percentage values of the normal distribution curve

Instead, from a consideration of Fig. 17 it is possible to comprehend the more significant characteristics of the normal curve in these terms.

In Fig. 17 is presented a normal curve symmetrically distributed about a mean value $\bar{x}$, the shape of this curve being expressed by the standard deviation of the set of data, i.e. σ. The values used are those mentioned at the beginning of this chapter, namely an average of 30 bushels per acre and a standard deviation of 5·0 bushels. The area between the curve and the base line is here divided by vertical lines which are drawn at a distance away from the average (both above and below it) equal to the standard deviation and to successive multiples of that deviation, e.g. at $\bar{x}$ plus 1σ which is $30 + 5 = 35$; then at $\bar{x}$ plus 2σ which is $30 + 10 = 40$ etc. By applying a rather complicated formula it is now possible to say what percentage of the whole set of data will lie between any successive pair of these vertical lines. The values which apply when the distribution curve is truly normal are entered on the diagram in Fig. 17 in a somewhat generalized form.

It can be seen that some two-thirds of the occurrences lie less than 1 standard deviation away from the average, i.e. between the values $\bar{x} - 1\sigma$ and $\bar{x} + 1\sigma$. Equally about 95% of the occurrences lie less than 2 standard deviations away from the average, while less than 1% of them differ from the average by more than 3 standard deviations. To be rather more precise, these values imply the following, provided that the curve is perfectly normal:

68·3% of the occurrences will lie between $+1\sigma$ and -1σ, i.e. there is roughly a 2 : 1 chance that a value will lie between those limits and a 1 : 2 chance that it will not.

95·45% of the occurrences will lie between $+2\sigma$ and -2σ, i.e. there is roughly a 21 : 1 chance that a value will lie between those limits and a 1 : 21 chance that it will not.

99·7% of the occurrences will lie between $+3\sigma$ and -3σ, i.e. there is roughly a 330 : 1 chance that a value will lie between those limits etc.

99·99% of the occurrences will lie between $+4\sigma$ and -4σ, i.e. there is only one chance in 10,000 that a value will differ from the average by more than this amount.

In a strict sense, a complete total of 100% is never reached, for the normal probability curve never cuts the horizontal axis, approaching it asymptotically instead.

The percentage values quoted in this way, or those shown in Fig. 17, are very useful as indicators of the scatter of actual values about the average, but they only provide figures for whole numbers of standard deviations. A more complete picture is obtained if more values are available. These have been calculated and are presented in print elsewhere. In Table IX a selection of these values is set out. This table gives the percentage of the occurrences which will lie within a certain number of standard deviations from the mean and also the number of standard deviations from the mean that will enclose certain percentages of the occurrences. Thus within + and $-2 \cdot 5\sigma$ lie 98·76% of the occurrences, while 50% of the occurrences will differ from the mean by *not more than* 0·6745 standard deviations.

Armed with this information it is now possible to look once more at the crop-yield example quoted at the beginning of this chapter. By reference to the percentage points of the normal distribution given in Table IX, it can readily be calculated that, for example, 80% of the occurrences lie between 23·59 and 36·41 bushels per acre (i.e. the

Table IX

Percentage points of the normal distribution

%	σ	%	σ
10	0·1257	90	1·6449
20	0·2533	92	1·7507
30	0·3853	94	1·8808
38·30	0·5000	95·45	2·0000
40	0·5244	96	2·0537
50	0·6745	98	2·3263
60	0·8416	98·76	2·5000
68·26	1·0000	99	2·5758
70	1·0364	99·73	3·0000
80	1·2816	99·95	3·5000
86·64	1·5000	99·99	4·0000

% = the percentage of the occurrences that will lie not more than the given number of σs away from the mean.
σ = the number of standard deviations away from the mean within which limits the given percentage of the occurrences will lie.

For full details see: D. V. Lindley and J. C. P. Miller, *Cambridge Elementary Statistical Tables*, Cambridge, 1953 (Table II).

mean $+/-1 \cdot 2816\sigma$), or that although the average value is 30 bushels per acre, individual values lie outside the range 27·5 to 32·5 bushels per acre on 61·7% of the times. The ability to assess such aspects with little further effort once the average and standard deviation are calculated represents one of the greatest advantages of working in terms of those units rather than any of the others which were considered in the previous two chapters.

In doing this, however, it is necessary to be sure that the distribution is reasonably normal. Given that this is so, the percentages quoted above will be found to hold true. This can be clearly illustrated from the Bidston rainfall data analysed previously. It was shown in Table III that the average for this set of data was 28·45″ while the standard deviation was 3·45″. From the percentage points of the normal distribution (Table IX) it can be seen that between $+1\sigma$ and -1σ, i.e. between 31·90″ and 25·00″, should lie 68·3% of the occurrences or a total of 20·5 occurrences. In reality, as reference to Table I where the values are arranged in order of magnitude will show, 21 of the 30 occurrences, or 70%, fell between these limits. This is as close to the calculated value of 68·3% as could possibly occur. Equally, between 35·35″ and 21·55″ (i.e. between + and -2σ) lie 29 of the occurrences which can be compared with the 28·6 occurrences (95·45%) which theory forecasts. In this particular set of data no value differs from the average by as much as 3 standard deviations. As such a difference should theoretically happen only 3 times in 1,000 and there are only 30 occurrences here being studied, this is not unexpected. Thus, as a concise statement of annual rainfall conditions at Bidston an average of 28·45″ and a standard deviation of 3·45″ tells a very great deal, and the addition of the standard deviation to the more usual information of the average vastly expands the information provided.

Three Standard Deviations Check

Furthermore, this example also illustrates another use of the standard deviation. It has been seen in the foregoing example that no value differs from the average by more than 3 standard deviations, largely because the number of occurrences is relatively few. Even with a large body of data a difference this great is only to be expected once in more than 300 occurrences. When assessing scatter by the standard deviation it is therefore useful to check the accuracy of the calculations and the data by seeing whether any record does differ from the average by more than 3σ. For example, if such a large difference from the average is found in a record of only 50 values, it is advisable to regard such a value with some suspicion. It may well be truly valid, for the exceptional case has to occur some time and it may be within the 50 occurrences being studied. On the other hand it is as well to check for errors. Perhaps a figure has been wrongly written or read; a small change in the character of the data may have been missed; a rain-gauge may have developed a leak! In other words, the set of data may not be strictly homogeneous nor the distribution curve approximately normal. This '3 *standard deviation' check* is therefore a safeguard against really gross errors. Thus it could well have been applied to the answer obtained by the grouped frequency calculation of the standard deviation of the 'number of poultry per farm' data set out in Table VII. Three times the standard deviation of 35·5 equals 106·5. As the average value was 100·04 and the range of values from 2 to 200 (p. 36), no value differs from the average by more than 3σ, despite the 1,044 occurrences. Clearly no major discrepancy would seem to be within the record, although this is no safeguard against minor ones. It could be argued equally, of course, that as it is not possible for values to be as much as three standard deviations below the average, then the frequency curve is not strictly normal (see p. 43).

Transformation of Data

In the previous sections it has been stressed many times not only that it is assumed that the data fit the normal frequency distribution but also that such an assumption underlies the larger part of statistical theory. Yet in Chapters 1 and 2 it was also seen that real bodies of data often do not satisfy this requirement, and that instead the data form a skew frequency distribution (Figs. 4 and 6). In these cases, the characteristics of the normal frequency distribution outlined above no longer strictly apply, which means that either one must be willing to forego the possibilities for further statistical evaluation of the data outlined in succeeding chapters, or the data must be modified or manipulated in such a way that the qualities of the normal frequency curve can still be utilized.

Such manipulation can be effected by a process termed *transformation*, by which the data are changed mathematically into a form that more closely approximates to the normal curve. Graphical examples of the transformation process are given in Fig. 18. In Case 1, the original data $X1$ display marked positive skewness, with a maximum frequency at the lower end of the scale between 75 and 100. Values fall to the 0 to 25 range, but rise continuously as high as 300 to 325,

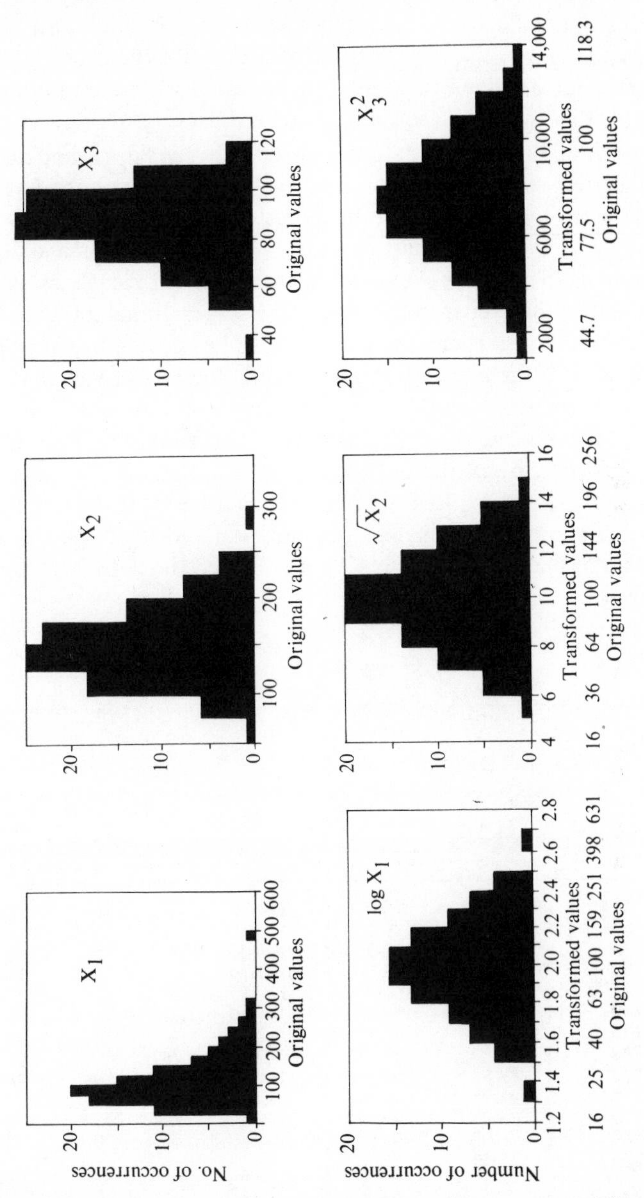

Figure 18. Examples of the transformation of skew data to a normal frequency distribution

with the isolated extreme occurrence at 475 to 500. If all these values are changed into their appropriate logarithms (to the base 10, in the present case), and then regraphed as log. *X*1, it will be seen that they now fit the normal frequency distribution very closely. Again, in Case 2 of Fig. 18 the data are also positively skew, though to a markedly lesser degree than in Case 1. This time a normal curve can be produced by transforming the data not into logarithms but into square roots instead. Finally, in Case 3 (Fig. 18) the original data *X*3 possess slight negative skewness, with maximum frequency occurring nearer the upper rather than the lower limits of the data. Transformation to normal in this case can be effected by calculating the squares of the original data and regraphing these values.

It is therefore essential that, if statistical techniques are to be used which assume a normal frequency distribution, the body of data must first be examined to assess whether or not such assumptions are reasonably justified. This can be done by using arithmetic probability paper, on which can be plotted the original data in only slightly modified form. In Fig. 19*a* are plotted the data from Case 1 of Fig. 18.

Table X

Preparation of data for use with probability paper (Fig. 19)

Original Data				*Transformed Data*			
Cells	Frequency	Cumulative Frequency or Rank	Cumulative Percentage using $\frac{m\text{-}0{\cdot}5}{n}\times 100\%$	Cells	Frequency	Cumulative Frequency or Rank	Cumulative Percentage using $\frac{m\text{-}0{\cdot}5}{n}\times 100\%$
1– 25	1	1	0·5	1·31–1·40	1	1	0·5
26– 50	11	12	11·5				
51– 75	18	30	29·5	1·51–1·60	4	5	4·5
76–100	20	50	49·5	1·61–1·70	7	12	11·5
101–125	15	65	64·5	1·71–1·80	9	21	20·5
126–150	11	76	75·5	1·81–1·90	13	34	33·5
151–175	7	83	82·5	1·91–2.00	16	50	49·5
176–200	5	88	87·5	2·01–2·10	16	66	65·5
201–225	4	92	91·5	2·11–2·20	13	79	78·5
226–250	3	95	94·5	2·21–2·30	9	88	87·5
251–275	2	97	96·5	2·31–2·40	7	95	94·5
276–300	1	98	97·5	2·41–2·50	4	99	98·5
301–325	1	99	98·5				
				2·61–2·70	1	100	99·5
476–500	1	100	99·5				

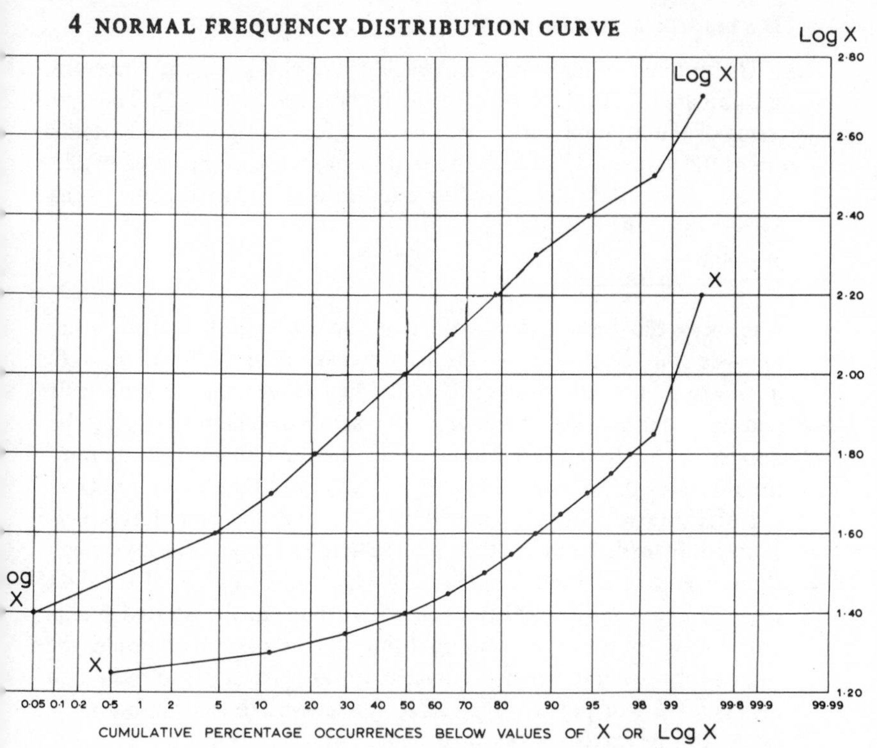

Figure 19*a*. Arithmetic probability paper Figure 19*b*. Logarithmic probability paper

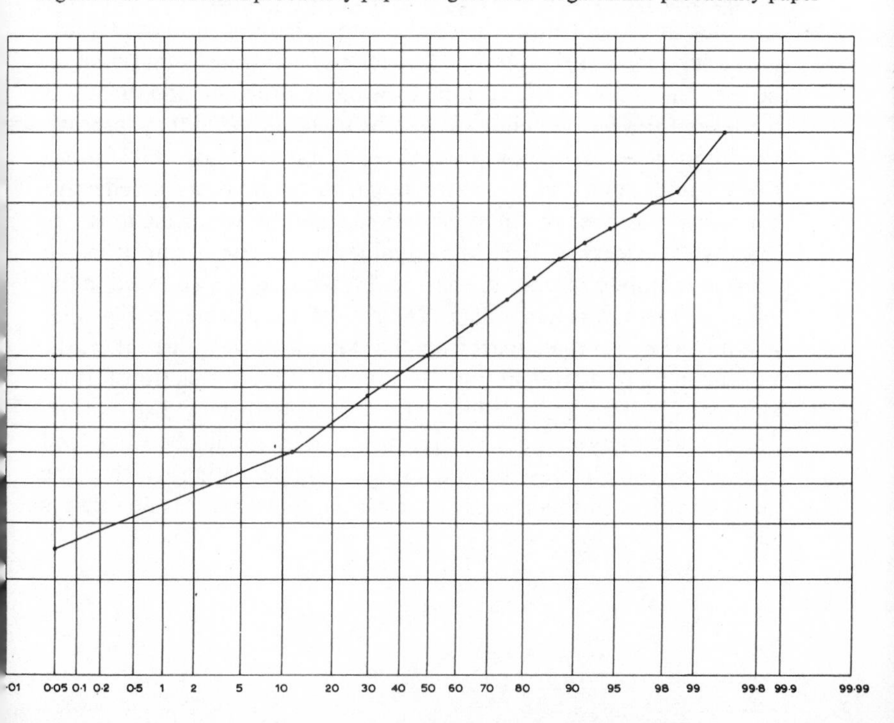

The number of occurrences in each cell is obtained, and these are accumulated from the lowest to the highest cell (Table X). Thus, in terms of the original data, it can be seen, for example, that 30 items are of the value of 75 or less, and that 92 items are of the value of 225 or less. These data are then converted into cumulative percentages by means of the formula:

$$\frac{m - 0{\cdot}5}{n} \times 100\%$$

when n = the total number of items (here $n = 100$) and m = the number of items at or below the given value (i.e. $m = 30$ and $m = 92$ in the two examples above). If individual values, rather than cells and cell frequencies, were used the same calculations would be applied for each item; in that case, m could be redefined as the rank number of a particular value and n as the total number of ranks.

Returning to Fig. 19*a*, it will be seen that the cumulative percentages from Table X are then plotted against the value below which the percentage of items falls, e.g. 64·5% are below 125. If the data approximate to a normal frequency distribution, the plotted points will fall on, or close to, a straight line. In the present example, the points form a curve which is concave upwards from left to right, and this is typical of positive skewness; if negative skewness applies, the curve will be convex upwards instead.

The positive skewness of the data thus being apparent, it is necessary to use the transformation process before most statistical analyses or tests are employed. To decide which transformation is to be used, however, is often a time-consuming problem, and there is a tendency for a few common ones to be used, provided they improve the frequency distribution, even if other less common transformations would, in fact, be yet better. Logarithms and roots are effective for most positive skewness situations, and power functions for negative skewness. If logarithmic values of $X1$ are obtained in the present example, as was done in Fig. 18, these can be analysed in the same way as the original data. The results are plotted on Fig. 19*a*, and it can be seen that, apart from the extreme upper values, these do approximate to a straight line, thus indicating that this transformation does lead to a closer fit to the normal frequency distribution. Ideally, a variety of possible transformations should be calculated and plotted, the closest fit to a straight line indicating which is the most effective. In this process of selection, however, it is not always

Table XI

Lead Production (thousands of metr c tons) 1961
(excluding China, N. Korea and U.S.S.R.)

	Original X	transformed log. $10X$
U.S.A.	744·6	3·8720
Australia	213·7	3·3298
Mexico	175·5	3·2442
Canada	155·9	3·1928
W. Germany	140·6	3·1479
Belgium	99·9	2·9996
Yugoslavia	90·4	2·9562
France	89·3	2·9509
U.K.	87·7	2·9430
Japan	83·3	2·9206
Peru	76·5	2·8837
Spain	66·3	2·8215
Italy	45·2	2·6551
Bulgaria	40·9	2·6117
	(40·35 —Median—	2·6058)
Poland	39·8	2·5999
Sweden	38·8	2·5888
Argentina	28·0	2·4472
Morocco	24·5	2·3892
Tunisia	18·5	2·2672
Burma	16·2	2·2095
Zambia	15·4	2·1875
Romania	12·0	2·0792
Netherlands	10·8	2·0334
Austria	6·5	1·8129
Brazil	4·9	1·6902
India	3·7	1·5682
Portugal	1·5	1·1761
Guatemala	0·1	0·0000
	83·2 —Average—	2·4849
	139·1—Standard deviation—	0·7561

Original Data

Classes	Incidences
720–759·9	1
200–239·9	1
160–199·9	1
120–159·9	2
80–119·9	5
40– 79·9	4
0– 39·9*	14

Transformed Data

Classes	Incidences
3·5–3·9999	1
3·0–3·4999	4
2·5–2·9999*	11
2·0–2·4999	7
1·5–1·9999	3
1·0–1·4999	1
0·5–0·9999	0
0·0–0·4999	1

*Modal class

necessary to carry out the actual transformation calculations, for the original data can again be plotted, this time upon, for example, logarithmic probability paper, as is illustrated in Fig. 19*b*. The basically straight line can again be seen, indicating that a logarithmic transformation of this body of data is essential.

Having thus arrived at a decision concerning which type of transformation to use in order to eliminate the influence of skewness, the original values must then all be converted in this way. A typical example from the field of economic production data is worth consideration. In Table XI are presented the production figures of lead, in thousands of metric tons, for most of the world's producers in 1961. It will be seen that the several expressions of the mean value for the original data are:

average = 83·2 median = 40·35 modal value = 23·3 (see formula on p. 12).

These suggest a markedly skew frequency distribution in a positive direction (p. 9). If these values are all converted into logarithms, this positive skewness is replaced by a slight negative skewness, and plotting of these two sets of values from Table XI on the probability paper of Fig. 19 will show a change from a concave to a convex curve. Thus the mean values for the logarithmic data are:

average = 2·4849 median = 2·6058 mode = 2·68

Furthermore, the reduction in skewness is shown by the relationship of the standard deviations to the size of the average values. Thus for the original data the standard deviation is 139·1 as compared to an average of 83·2, while with the transformed data the standard deviation and average values are 0·7561 and 2·4849 respectively. (Note in this connection that all the original values were multiplied by 10 before the logarithms were found, to eliminate the possibility of negative values).

It is with values derived from these transformed data that any further analyses are concerned, when techniques are to be used that employ relationships derived from the qualities of the normal curve. This applies in such fields as probability assessments (Chapter 5), significance testing of results by parametric techniques (Chapters 8 and 9), and correlation and certain aspects of regression (Chapters 11 and 12). It must always be remembered, however, that in interpreting the results of such analyses in terms of the real problem with which one is concerned, values and results must be retransformed back into the form of the original data.

Skewness and Kurtosis Indices

Although most of the attention in this and later chapters is upon the mean and the standard deviation as characteristics of the normal frequency distribution, there are two other characteristics which also require mention. These are skewness and kurtosis.

As has been seen earlier, *skewness* is the term used to describe the extent to which the frequency curve is symmetrical or asymmetrical, together with the direction of asymmetry. The basic expression of the formula for the skewness index is the 'third moment about the mean' (i.e. the average of the sum of the cubes of the deviations) divided by the standard deviation cubed. Thus:

$$\text{skewness} = \frac{\dfrac{\sum (x-\bar{x})^3}{n}}{\left(\sqrt{\dfrac{\sum (x-\bar{x})^2}{n}}\right)^3}$$

With a normal curve this value is zero, values increasing as asymmetry increases. Moreover, a positive index is obtained if the peak frequency lies in the low-value ranges (Fig. 4), and a negative index if the peak frequency is in the high-value ranges (Fig. 6)—hence the terms positive and negative skewness.

The term *kurtosis* applies to the degree to which the frequency distribution is concentrated around the frequency peak, i.e. it describes the degree of 'peakedness' of the curve. The basic formula for this is to divide the 'fourth moment' (i.e. the average of the sum of the fourth powers of the deviations) by the standard deviation raised to the fourth power. Thus:

$$\text{Kurtosis} = \frac{\dfrac{\sum (x-\bar{x})^4}{n}}{\left(\sqrt{\dfrac{\sum (x-\bar{x})^2}{n}}\right)^4}$$

For a perfectly normal distribution this yields a value of 3. An index less than this indicates a lesser concentration about the mean than would apply with a normal curve, while an index greater than 3 indicates a greater-than-normal degree of 'peakedness'.

CHAPTER 5

PROBABILITY ASSESSMENTS

Probability Theory

In the discussion of the normal frequency distribution mention has several times been made of the percentage of occurrences within certain limits, or of the chance of certain values occurring. This introduces a theme which is fundamental to the whole of statistical analysis, namely the theme of probability. In the case of a large proportion of analyses one of the main purposes is to assess the probability that given values are likely to occur, or to be exceeded, or not to be reached. From other points of view the problem may also be posed in a rather different way although it is still basically the same problem. Thus the question may be asked as to the probability that certain events are likely to occur at given intervals, or that a certain distribution pattern has some significant meaning. Moreover, even to interpret the average properly it is necessary to think in terms of probability—the probability of its being exceeded or not, for example.

This field of probability theory is vast and complex in detail, although its fundamentals are simple enough. If the full set of any body of data is considered, the probability that any individual occurrence will lie between the values for the outer limits of that complete set is obviously 100%. Equally, the probability of any value being equal to or lower than the highest value of the set is also 100%, as is the probability of any value being equal to or greater than the lowest value of the set. In other words, if the full set of data is considered it must necessarily contain all the events and therefore all the probabilities. The fact that the full set of data represents 100% probability is shown by expressing this total probability as unity, i.e. as 1·0.

If, on the other hand, the probabilities of values being greater than average and less than average were to be considered then, assuming a perfectly normal frequency distribution, each of these events would occur with a 50% probability, i.e. there would be an equal likelihood of a value being above or below the average, and a complete certainty that it must be one or the other. This can be tabulated as follows:

probability of a value being greater than average	=	50% or 0·5
,, ,, ,, ,, less ,, ,,	=	50% or 0·5
total probability of the value being greater or less than the average	=	100% or 1·0

Again, by reference to Table IX it can be seen that the following probabilities hold true, if the distribution is normal:

probability of a value differing from the average by less than 2σ	=	95·45% or 0·9545
probability of a value differing from the average by more than 2σ	=	4·55% or 0·0455
total probability of a value differing from the average by more or less than 2σ	=	100% or 1·0

These two simple examples make it clear that the sum of the individual probabilities within a set of data is the same as the total probability which is unity.

The problem of assessing the probability with which given values or events are likely to occur is thus basically a problem of deciding how to allocate the total probability between the various possibilities under review. In the above examples only two possibilities were present in each case, but far more complex conditions can be argued in the same way. It must be realized and accepted from the outset, however, that a statement of probabilities in this way does *not* indicate *when* the specified conditions will occur. It does no more than assess the frequency with which those conditions are likely to occur over an *infinitely long* set of records. The longer that set may be, the closer actual frequencies or probabilities are likely to be to these theoretical values. This theme will be expanded at greater length when the taking and analysing of samples, and the question of their size, are considered in Chapter 6.

The problem mentioned above of allocating the total probability between various possibilities must be decided in terms of the type of frequency distribution curve which most closely fits, or approximates to, the curve of the data itself. For many sets of data it is the normal curve which is the relevant one. For other sets of data, however, or for other purposes, rather different distribution curves may apply.

Of these the most common and most useful are the binomial distribution and the Poisson distribution. These two will therefore be considered and illustrated after the normal curve and its implications have been further examined. These various possibilities of assessing probability will be simply presented with a minimum of mathematical theory and a maximum of practical value.

Probability and the Normal Frequency Distribution

From the consideration of the percentage points of the normal distribution earlier in Chapter 4 several indications were given concerning the probability with which specified conditions occur. Thus it was seen that the probability of a value differing from the average by more than 2σ was 4·55%, or, in terms of annual rainfall at Bidston, that values outside the range 21·55″ to 35·35″ will occur with this same probability or frequency. Very often, however, it is not the probability of values falling within a certain range which is relevant and of interest but rather the probability that values will exceed or fall below some given value. For example, it may be of value to know the probability that an occurrence will exceed the average by more than 2σ, or that rainfall at Bidston will be greater than 35·35″ (which is itself 2σ greater than the average). Clearly, if the distribution is a normal one, the 4·55% probability that a value will differ from the average by more than 2σ will be equally distributed between the two ends of the curve, i.e. between values greater than $\bar{x} + 2\sigma$ and values less than $\bar{x} - 2\sigma$. This has, in fact, already been shown diagrammatically in Fig. 17. Therefore, having obtained from Table IX the fact that 95·45% of the values lie between -2σ and $+2\sigma$, and that therefore 4·55% fall outside these limits, it is simply a matter of halving this latter value to find the percentage of values that are likely to be greater than $\bar{x} + 2\sigma$. So it can be established that 2·275% of the values should fall into this category, or in terms of annual rainfall at Bidston that in 2·275% of the years rainfall should be greater than 35·35″. This one chance in 40 does, in fact, occur once in the thirty-year record set out in Table I, but the similar probability of a fall of less than 21·55″ does not occur within that particular short period.

This reasoning has been presented at some length because it is basic to the calculation of probabilities of any and every value. Table IX referred to above, from which was obtained the percentage probability of values being between $+2\sigma$ and -2σ, is not in the most convenient form for other calculations. Therefore, if intermediate values of standard deviations are required, or if the problem is posed in terms of the probability of a given value being exceeded, a simple calculation and reference to tables of the *normal distribution function* can be made. The calculation is as follows:

$$\text{required figure} = \frac{\text{critical value} - \text{mean value}}{\text{standard deviation}}$$

which is usually written $d = \dfrac{x - \bar{x}}{\sigma}$

The required figure or d is the figure which is needed for reference to tables. Into the right-hand side of the formula can be entered the mean and standard deviation values, and also the value which is being investigated. The calculation gives an answer which indicates the extent to which the critical value differs from the mean expressed in terms of 'so many' standard deviations. Thus, to recalculate the earlier Bidston example by this method, the following would be done. Suppose that it is desired to know the percentage probability that values will exceed 35·35″ of rainfall, this being then the critical value in the formula. Values can be entered in this way:

$$d = \frac{x - \bar{x}}{\sigma} = \frac{35{\cdot}35 - 28{\cdot}45}{3{\cdot}45} = \frac{+6{\cdot}90}{3{\cdot}45} = +2{\cdot}0$$

From this required figure of $d = 2{\cdot}0$ the appropriate percentage probability is then obtained from Table XII, the Normal Distribution Function. The value in this case is 2·275%, and since d is positive this indicates the percentage probability that occurrences will be *more than* the critical value. This is the same probability as that obtained by the alternative method on p. 62. The probability of occurrences being *less than* this value is obtained by '100 — tabled percentage', i.e. 100 — 2·275 which equals 97·725%. Conversely, if it were desired to know the probability of occurrences below 21·55″ a similar calculation would be made:

$$d = \frac{x - \bar{x}}{\sigma} = \frac{21{\cdot}55 - 28{\cdot}45}{3{\cdot}45} = \frac{-6{\cdot}90}{3{\cdot}45} = -2{\cdot}0$$

Table XII

The Normal Distribution Function

d	%	*d*	%	*d*	%	*d*	%	*d*	%
0·00	50·00	0·50	30.85	1·00	15·87	1·50	6·68	2·0	2·275
0·10	46·02	0·60	27·43	1·10	13·57	1·60	5·48	2·5	0·621
0·20	42·07	0·70	24·20	1·20	11·51	1·70	4·46	3·0	0·135
0·30	38·21	0·80	21·19	1·30	9·68	1·80	3·59	3·5	0·023
0·40	34·46	0·90	18·41	1·40	8·08	1·90	2·87	4·0	0·003

If 'd' is positive

d = the number of standard deviations that the critical value is *above* the mean.

% = the percentage probability that the occurrence will be *more than* the corresponding value of '*d*'; the probability that it will be *less than* this value is (100 – %).

If 'd' is negative

d = the number of standard deviations that the critical value is *below* the mean.

% = the percentage probability that the occurrence will be *less than* the corresponding value of '*d*'; the probability that it will be *more than* this value is (100 – %).

For a more detailed table see D. V. Lindley and J. C. P. Miller, *Cambridge Elementary Statistical Tables*, Cambridge, 1953 (Table I).

Again Table XII may be used, but because the *d* value in the above calculation is negative the percentage values have to be interpreted in the reverse way, as is indicated in the footnote to the table itself. Interpreted in this way the table gives the percentage probability of values being *less than* the critical value, and the adjustment by means of '100 – tabled percentage' is used to obtain the probability of occurrences above the critical value. So in the present example the percentage probability of annual rainfall at Bidston being below 21·55″ is again 2·275%. This dual use of the table is necessitated by the fact that it gives values along only one side of the distribution curve, i.e. between the average and any *one* end of the curve. This is because it is concerned with the probability of values being exceeded or not, rather than with the probability of values falling within certain limits, as was Table IX. These two tables (IX and XII) are, of course, simply different ways of expressing the same set of relationships, both being based on the form of the normal frequency curve described earlier.

In this example the results of the probability assessments have been compared to the conditions within the data used, simply to indicate the general reliability of the results. Strictly speaking, however, the analysis is carried out on sample data, and the results appertain to the total population. This means, for example, that such studies of sample rainfall data can provide valuable information in relation to water-supply problems, irrigation requirements, river run-off and

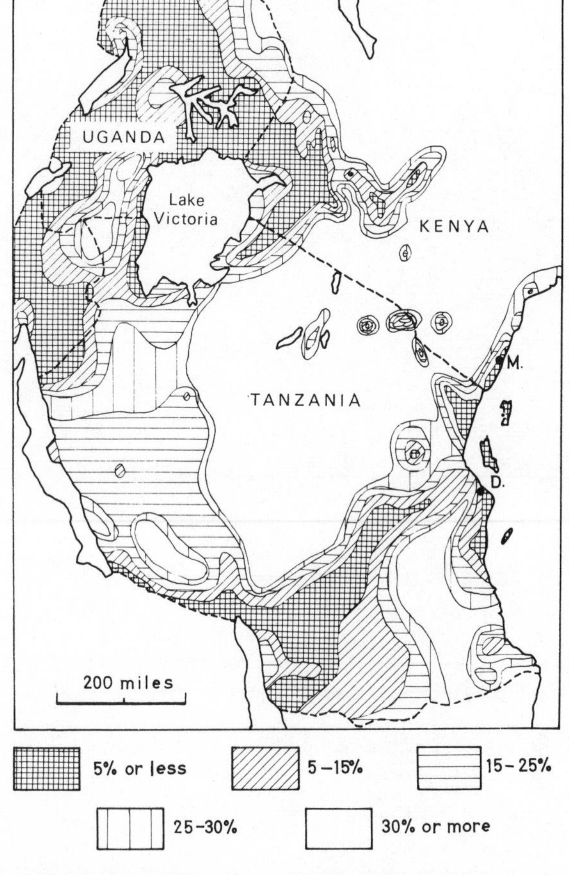

Figure 20. Rainfall probability map of East Africa (after J. Glover, P. Robinson, J. P. Henderson, *Quart. J. R. Met. Soc.*, 80 (1954))

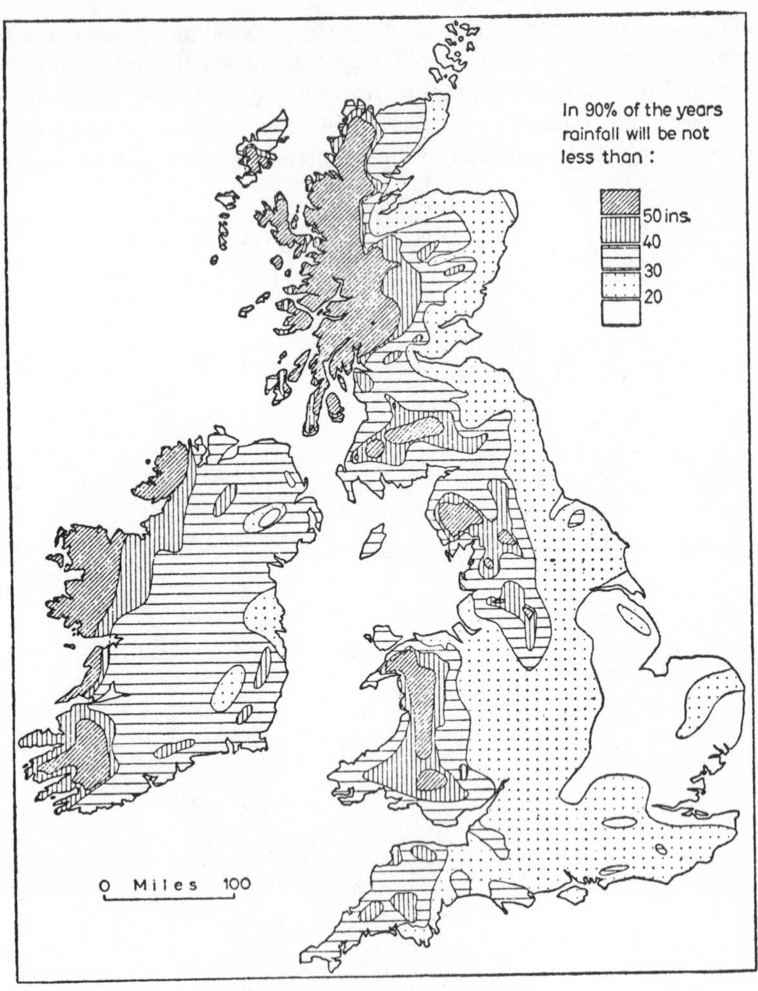

Figure 21. Rainfall probability map of the British Isles

flood conditions over a time-scale far greater than that for which data are available. Maps showing rainfall probability characteristics have been prepared for various countries, and Figs. 20 and 21 provide two examples.

The implications and possibilities of this method will be more fully appreciated, however, if another set of data is analysed and various probability features assessed. For this purpose it is convenient to use the sample data concerning the number of poultry per farm presented in Table VII and examined in Chapter 3. Use of these data has the following advantages: the mean and standard deviation values are already calculated and the distribution curve has been seen to be very close to normal. The arithmetic average of this set of data was 100·04 and the standard deviation 35·5. Suppose that only these values were known, and the original data were not available. It may be desired, nevertheless, to assess the probability of occurrence of certain conditions.

The first inquiry is to discover the percentage probability that more than 140 head of poultry will occur on a farm—or, to put it another way, the percentage of all the farms in the area (not simply the sample farms) which are likely to have more than 140 head of poultry. This can be readily calculated as in the previous example, thus:

$$d = \frac{x - \bar{x}}{\sigma} = \frac{\text{critical value} - \text{mean value}}{\text{standard deviation}}$$

$$= \frac{140 - 100{\cdot}04}{35{\cdot}5} = \frac{+39{\cdot}96}{35{\cdot}5} = +1{\cdot}125$$

In this case the *d* value is positive, so that the percentage probability of exceeding the critical value can be read directly from Table XII. This indicates that more than 140 poultry can be expected to be found on 13·03% of the farms.

Again, it may be desired to find out how many farms have relatively few poultry, taking 20 as the critical value. The calculations follow the same line as above:

$$d = \frac{x - \bar{x}}{\sigma}$$

$$= \frac{20 - 100{\cdot}04}{35{\cdot}5} = \frac{-79{\cdot}96}{35{\cdot}5} = -2{\cdot}25$$

Here the *d* value is negative, but as it is the proportion of occurrences *below* this value that is required the necessary value can again be read directly from Table XII. With the abbreviated version given here it can only be placed between 0·621% and 2·275% probability, but from the full tables the correct value is seen to be 1·22%. It is this ability to infer from *sample data* the probable frequency of occurrence of specified conditions within the *total population* which forms one of the major assets of these types of analysis.

Apart from discovering the probability with which given values can be expected to be exceeded it is also often desirable to assess the value that can be expected to occur or be exceeded with a given probability. For example, it could well be of interest to define the number of poultry which is equalled or exceeded on 80% of the farms. Probability values are tabulated in terms of half the distribution curve, however, so that it is convenient to pose the problem in terms of one half of the curve or the other, rather than in terms of something that overlaps the mean value. Thus, this problem could be put as one of defining that value below which will fall 20% of the occurrences. The value must therefore of necessity be below the mean and the d value will be a negative one. What will it be? In this case it is possible to know this value in advance from the normal curve, for it is the value of x (the critical value) which is to be discovered. To obtain the d value it is necessary to consult Table XII again to find the value which will ensure that 20% of the occurrences fall below it. This is seen to lie between 0·80 and 0·90, and detailed tables give it as 0·8416. In other words, 20% of the occurrences lie more than 0·8416σ below the average, while 80% of the occurrences lie above this value, i.e. this is the value that the problem is concerned with. Now it is possible to insert all but one of the values into the probability formula. Thus:

$$d = \frac{x - \bar{x}}{\sigma}$$

$$d.\sigma = x - \bar{x}$$

$$\bar{x} + d.\sigma = x$$

i.e. $100{\cdot}04 + (-0{\cdot}8416 \times 35{\cdot}5) = x$

$100{\cdot}04 - 29{\cdot}88 = x = \underline{\underline{70{\cdot}16}}$

Thus 80% of the farms in an infinite series will possess more than 70·16 poultry (i.e. 71 or more), while 20% of the farms will possess less than this amount. This formula for assessing such values is simply an adjustment of the one presented earlier, and can be put in a standard form as follows:

critical value = d(standard deviation) + the mean

or $x = d.\sigma + \bar{x}$

always bearing in mind that *d* may be a negative value as in the above example. The relationships implied by these two forms of the formula are presented diagrammatically in Fig. 22.

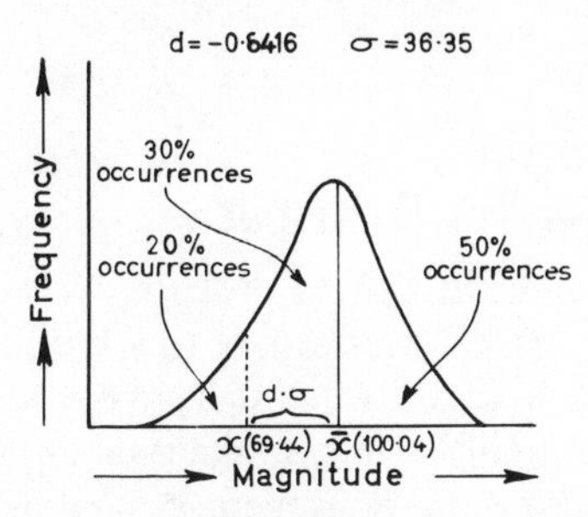

Figure 22. Diagram of calculating probability values from the normal distribution curve

Whilst calculations such as these are often necessary, it is also possible on many occasions to make probability assessments by using probability paper instead. At the end of Chapter 4 the use of such paper was outlined for the assessment of whether or not a body of data approximated to the normal frequency distribution, and the efficacy of transformation techniques in producing such a distribution. Given that a virtually normal curve is thus established by the production of a straight-line graph, it can then be used for probability assessments without any further calculations. For example, in Fig. 19*b* the transformed data are shown to approximate to a normal frequency distribution. Therefore, if it is desired to assess the probability of a value of below 70 occurring, this can be directly abstracted from the graph by reading the appropriate percentage value against the *x* value of 70. Thus 26% of the occurrences are likely to lie below 70, and equally 74% of the values are likely to be above this point. Again, the proportion of the values likely to lie between, say, 70 and 90 may be needed. By reference to Figure 19*b* it will be seen that 42% of the values are below 90; thus between 90 and 70 lie (42–26)% of the values, i.e. 16%. Conversely, to obtain the value which is exceeded 90% of the time, i.e. below which are only 10% of the occurrences, the appropriate value against the 10% line can be obtained. This is seen to be c. 48·5. This clearly establishes the practical, as well as the theoretical, value of plotting the cumulative data on probability paper, both to establish that techniques based on the normal curve are appropriate and to enable probability assessments to be made without any further calculations being required.

Probability and the Binomial Frequency Distribution

In all these considerations the probability values obtained have specifically omitted any suggestion as to *when* the stated conditions might be expected to occur, while it has been stressed (pp. 61 and 68) that such values apply strictly to the total population rather than to a limited part of the data. Thus there is no implication that because a given value is exceeded with an 80% probability that in any 10 occurrences 8 of them will be above the given value, although in 10,000 occurrences it is likely that approximately 8,000 will exceed it. Even less has any suggestion been made as to *which* of any 10 occurrences are likely to exceed that value, and which drop below it. This falls into the realms of forecasting, not of statistical probability. What can be attempted by means of statistical analysis, however, is to indicate the probability that in any ten occurrences 0, 1, 2, 3, 4, 5, 6, 7, 8, 9, or 10 of them will fall into the category of exceeding the given value. To be able to do this has obvious practical implications in terms of the reliance that can be placed on conclusions drawn from certain amounts of data, or the number of occurrences that need to be considered before an adequate degree of reliability can be obtained—a theme to be taken up more fully in Chapters 6 and 7. Also, in terms of interpreting data the probabilities of certain conditions occurring with a particular frequency may be far more valuable than a simple use of the mean or even of the overall probabilities already considered.

To obtain such probability values involves the consideration of another distribution curve, namely the *binomial distribution*. This is concerned with the relative frequency of occurrence of *two* numbers, or rather sets of *conditions*, which are mutually exclusive and which together represent the sum total of probability. Thus once a given set of conditions or a value is accepted as being critical and therefore worth analysing, then all the occurrences in the body of data can be classified as either belonging to that set of conditions or as not so belonging. This gives the overall long-term probability of these conditions occurring, either by counting in a sample body of data or by assessment from the normal distribution curve for an infinite body of data. Given that some specified number of occurrences are to be considered, it is possible for all these occurrences to belong to that set of conditions *or* for none of them to belong to that set of conditions *or* for some to belong and some not to belong, the proportions of each being liable to as many differences as there are occurrences under study. The prime characteristic of the binomial distribution is that it reflects the frequency (or the probability) with which these different possibilities are likely to occur, for any given percentage probability of the specified conditions and any given number of occurrences being considered.

A simple illustration may help to make the general principle clear before actual examples are analysed. Assuming that the data under consideration are normally distributed, what is the probability that in choosing *two* occurrences both will be above average *or* that both will be below average, *or* that there will be one above and one below average? In this case the number of occurrences being considered is two, while the specified set of conditions is that the value is above average. The overall probability of an above-average value is 50% or 0·5, as a normal distribution is assumed. Equally, the probability that a value will not be above average, i.e. will be below average, is also 0·5. From these data it is now possible to assess the probabilities sought at the beginning of this example. In a simple case such as this it can be done by tabulating all the possible combinations.

	First possibility	Second possibility	Third possibility	Fourth possibility
Above average	1. 2.	1.	2.	—
Below average	—	2.	1.	1. 2.

Thus both occurrences could be above average; both could be below average; and there are *two* ways in which one above average and one below average value could occur. In other words out of four possible combinations, only *one* could give both occurrences above average, i.e. the probability of this happening is 0·25. This is also true for both values below average, while there is a 0·5 probability of one of each of the two categories occurring. If this example is now turned from numbers into symbols the means by which these probabilities are obtained will be seen. Thus the specified above-average conditions can be called p, and those occurrences that do not satisfy these conditions can be called q, the data being retabulated in this form.

	First poss.		Second poss.		Third poss.		Fourth poss.	
Symbols:	p	p	p	q	q	p	q	q
Individual probability:	0·5	0·5	0·5	0·5	0·5	0·5	0·5	0·5
Overall probability:	0·25		0·25		0·25		0·25	
	$= 0{\cdot}5 \times 0{\cdot}5$		$= 0{\cdot}5 \times 0{\cdot}5$		$= 0{\cdot}5 \times 0{\cdot}5$		$= 0{\cdot}5 \times 0{\cdot}5$	
	$= p \times p$		$= p \times q = pq$		$= q \times p = pq$		$= q \times q$	
	$= p^2$		$= 2pq$				$= q^2$	

Thus the two probabilities of 0·25 and the one of 0·5 are seen to result from the multiplication of the individual probabilities, this 'Multiplication Law' applying in the case of the simultaneous occurrence of events as well as for assessing the probability of events in succession (see p. 176). The essential 'rightness' of this process and of the results is clear in the tabulation. Moreover, the setting in succession of the terms p^2, $2pq$ and q^2 should recall certain aspects of simple algebra acquired at the age of twelve or thirteen, for $p^2 + 2pq + q^2$ is the expansion of $(p + q)^2$. In other words, the probabilities of getting 2 occurrences of p, 1 occurrence of each of p and q, and 2 occurrences of q are given by the terms of the expansion of $(p + q)^2$. Furthermore, the power to which $(p + q)$ is raised, i.e. 2, equates with the number of occurrences being considered, i.e. 2, and it can be shown that the same relationship holds true whatever number of occurrences are being considered. Therefore the general formula for obtaining the individual terms of the binomial distribution is written as $(p + q)^n$, the expansion of this yielding the successive probabilities from all occurrences of p to all occurrences of q.

This is applied in the following way. In a given set of data it is known that the proportion with characteristic p is 0·2 so that the proportion without this characteristic, i.e. q, is 0·8. It is required to know the different probabilities of the various possible combinations of p and q, if 5 occurrences are being examined. The basic formula $(p + q)^n$ thus becomes $(0{\cdot}2 + 0{\cdot}8)^5$, or in its expanded form

$$p^5 + 5p^4q + 10p^3q^2 + 10p^2q^3 + 5pq^4 + q^5$$

Inserting the appropriate numerical values this becomes

$$0{\cdot}0003 + 0{\cdot}0064 + 0{\cdot}0512 + 0{\cdot}2047 + 0{\cdot}4097 + 0{\cdot}3277$$

These are then allocated as follows:

probability	of 5	occurrences	of p and 0 of q	= 0·0003
,,	,, 4	,,	,, p and 1 of q	= 0·0064
,,	,, 3	,,	,, p and 2 of q	= 0·0512
,,	,, 2	,,	,, p and 3 of q	= 0·2047
,,	,, 1	,,	,, p and 4 of q	= 0·4097
,,	,, 0	,,	,, p and 5 of q	= 0·3277
			Total probability	= 1·0000

The definition of the various terms applying to the different frequencies may well raise problems. The first such problem will probably be to establish the powers to which p and q must be raised for the different terms. Again working from all the occurrences being p to all the occurrences being q, i.e. from left to right in the equation on p. 72, in the first case the power of p is equal to n and that of q is nil. The power for the former steadily decreases by one each time moving from left to right while that of q equally steadily increases from nil to n in the same direction. This can therefore be written as follows:

$p^n; p^{n-1}q; p^{n-2}q^2$ etc. to $p^2q^{n-2}; pq^{n-1}; q^n$.

The other possible problem in the use of this technique is to assess the number of times by which the various values of p and q must be multiplied, i.e. the appropriate coefficients. These can be obtained without calculation by the use of what is known as 'Pascal's Triangle', which is set out in Table XIII. The values in this table can be extended simply beyond $n = 10$ by the process of addition. Thus, line $n = 4$ is obtained from line $n = 3$ by adding each successive pair of values in line $n = 3$ together, i.e. $0 + 1 = 1$; $1 + 3 = 4$; $3 + 3 = 6$; $3 + 1 = 4$; $1 + 0 = 1$; in this way the coefficients when $n = 4$ are seen to be 1, 4, 6, 4, 1.

Thus, if there were 8 occurrences, i.e. $n = 8$, then the terms of the expansion of $(p + q)^8$ would be:

$$p^8 + 8p^7q + 28p^6q^2 + 56p^5q^3 + 70p^4q^4 + 56p^3q^5 + 28p^2q^6 + 8pq^7 + q^8$$

This gives the full range of probabilities from eight occurrences of the given conditions p to no occurrences of these conditions but eight occurrences of the reverse conditions q instead.

Table XIII

Pascal's Triangle

Number in the sample $= n$	Coefficients in the expansion of $(p + q)^n$
1	1 1
2	1 2 1
3	1 3 3 1
4	1 4 6 4 1
5	1 5 10 10 5 1
6	1 6 15 20 15 6 1
7	1 7 21 35 35 21 7 1
8	1 8 28 56 70 56 28 8 1
9	1 9 36 84 126 126 84 36 9 1
10	1 10 45 120 210 252 210 120 45 10 1

It will be apparent that in all cases the first and last coefficients (for terms p^n and q^n) are unity; equally, for the second and the penultimate terms (i.e. for terms $p^{n-1}q$ and pq^{n-1}) the coefficients are always equal to n. These are perhaps the most commonly required terms, but if intervening terms are required when n is a large number, it may be convenient to use something other than an expanded Pascal's Triangle. So, for all except the first and last terms, the following equation can be used for the coefficient:

$$\text{coefficient} = \frac{n!}{r!\,(n-r)!}$$

when $n =$ the number of occurrences being considered; $r =$ the particular term in the expansion (starting from the left) minus 1; ! indicates that it is the 'factorial' of the number concerned, i.e.

$3! = 3 \times 2 \times 1 = 6$

$6! = 6 \times 5 \times 4 \times 3 \times 2 \times 1 = 720$

(for large numbers these can be obtained from prepared tables.)

Thus, if $n = 5$, the following coefficients obtain

1st coefficient = unity

$$\text{2nd coefficient} = \frac{5!}{1!\ 4!} = \frac{120}{1 \times 24} = 5 \text{ (i.e. } = n)$$

$$\text{3rd coefficient} = \frac{5!}{2!\ 3!} = \frac{120}{2 \times 6} = 10$$

$$\text{4th coefficient} = \frac{5!}{3!\ 2!} = \frac{120}{6 \times 2} = 10$$

$$\text{5th coefficient} = \frac{5!}{4!\ 1!} = \frac{120}{24 \times 1} = 5 \text{ (i.e. } = n)$$

6th coefficient = unity

A series of practical examples will illustrate this method in various ways and will also present several of the sorts of geographical problems that can be tackled by the use of this method of analysis. Suppose, for example, that it were known that in a particular area an annual rainfall of less than 20″ would result in a very poor harvest and furthermore that two such years in succession would lead to many farmers becoming bankrupt, much land going out of cultivation and the danger of famine. An analysis of the rainfall records indicates that a rainfall of below 20″ is likely to occur with a 10% probability, i.e. that there is a 10% chance of such a low value occurring or that on average it is likely to occur 1 year in 10. One such year can be survived, albeit with difficulty, and the problem therefore resolves itself into an assessment of the probability of two such years occurring in succession. This question can be analysed by means of the binomial distribution, for the probability with which the given conditions will occur is known to be 0·1, and the number of occurrences under consideration is 2. Thus into the formula $(p + q)^n$ can be entered the values

$p = 0{\cdot}1$ i.e. a 10% probability of receiving the given conditions;

$q = 0{\cdot}9$ i.e. a 90% probability that these given conditions will not be received and rainfall will be above 20″;

$n = 2$ i.e. the probabilities of receiving p and q in 2 successive years is required.

The expansion of these terms can be obtained in the way shown earlier and can be set out as follows:

Conditions	Calculations	Probability
Both years below 20″	$= p^2 = 0{\cdot}1^2$	$= 0{\cdot}01$
One year below 20″ and one year above	$= 2pq = 2 \times 0{\cdot}1 \times 0{\cdot}9$	$= 0{\cdot}18$
Both years above 20″	$= q^2 = 0{\cdot}9^2$	$= 0{\cdot}81$
	Total probability	$= 1{\cdot}00$

Thus it can be seen that with the conditions that were specified above, which were based on both the mean and the standard deviation parameters to obtain the percentage probability value for a year with below 20″ rainfall, two successive years with this low rainfall will occur with a probability of 0·01, i.e. there is a 1% chance of its occurring. Equally it shows that out of any pair of years there is an 18% chance that *one* of them will be dry, while there is an 81% probability that both years will be above the critical value. Values such as these may be markedly different from those which are often assumed from the study of mean values alone, or even from the more detailed results of variability analysis. In this case it means that conditions leading to famine, i.e. two successive dry years, will occur very infrequently despite the occurrence of such dry conditions in 10% of all the years.

This same method of analysis can, of course, be used in many other problems. For example, a given place may have an average long-term temperature for its warmest month of 65°F, which may be adequate for the maintenance of growth for certain trees. Such a temperature may not, however, be warm enough for the fruiting and regeneration of such trees, for which a mean temperature for the warmest month of 72°F may be required. With a life-span for the trees of about 100 years, such conditions are only essential at least once a century, to ensure replacement as old trees die out. By considering the standard deviation of the temperature data it is possible to discover the *overall* frequency with which such warmer conditions occur. If the standard deviation were found to be 3°F this would mean that

$$d = \frac{x - \bar{x}}{\sigma} = \frac{72 - 65}{3} = \frac{7}{3} = +2{\cdot}33$$

and from the normal distribution function (Table XII) this implies that the critical value, i.e. 72°F, is exceeded on 1% of the occurrences. In terms of an infinitely-long series of data the necessary warmth occurs with *just* the minimum frequency to ensure regeneration. There is no guarantee, however, that because the overall percentage probability is 1% that these conditions will occur with this frequency regularly, i.e. once every hundred years. The likelihood of temperatures above 72°F occurring with given frequencies within a period of a hundred years can be assessed by the binomial distribution, however. In this case the components of the formula $(p + q)^n$ are:

$p = 0{\cdot}01$ this being the probability of receiving a mean temperature for the month above 72°F;

$q = 0{\cdot}99$ this being the probability of not receiving more than that amount;

$n = 100$ this being the critical period within which it is necessary for this temperature to be received.

What is now wanted is the probability of a monthly temperature of the warmest month being over 72°F occurring some time during a hundred years. As the total probability of values for differing proportions of 'above and below 72°F' must equal unity, the simplest way to obtain the required answer is to calculate the probability that *no* year with a monthly temperature above 72°F will occur within the hundred years; subtracting this from unity will give the probability value required. The probability that there will be no values of p is obtained by calculating q^n, which is the last of the terms of the expansion of $(p + q)^n$ (see p. 74). Thus, the probability of no p value $= q^n = 0{\cdot}99^{100} = 0{\cdot}366$. Therefore, the probability of some p values $= 1 - 0{\cdot}366 = 0{\cdot}634$.

It can therefore be seen that although there is a more than 60% probability that one or more years in a hundred will experience temperatures adequate for tree regeneration, there is a 35 to 40% probability that not even *one* year out of the hundred will receive such adequate temperatures. It would thus appear that it is quite possible for trees to fail to regenerate under these conditions, after possibly several centuries of continued existence and regeneration, without any real change in climate tc account for this change in vegetation. The 'change' which would have occurred would be no more than the random occurrence of exceptionally warm conditions with an overall frequency of 1%, this necessarily implying that at times a period of more than a hundred years will lapse between such occurrences, while at other times these occurrences will be slightly more frequent for several centuries. It must not be assumed that the above argument proves or disproves changes in climate. As presented here it is simply an example of a possible set of relationships, but it does indicate the type of problem that may well repay analysis by this method.

A final example may reinforce the understanding of these methods. Suppose, for example, that in a given area it is known that 60% of the farms include dairying within their economy. In a brief visit, perhaps on a field excursion, it proves possible to visit three farms within this area. What are the probabilities of these visits including 3, 2, 1 or even 0 farms with dairying activities? The components p, q and n can be set out as before:

p (the proportion with dairying) $= 0{\cdot}6$
q („ „ without „) $= 0{\cdot}4$
n (the number of farms being visited) $= 3$

The proportions are as follows, still following the working principles set out on p. 74.

The probability of 3 farms with dairying $= p^3 = 0{\cdot}6^3 = 0{\cdot}216$
„ „ „ 2 „ „ „ $= 3p^2q$
$= 3 \times 0{\cdot}6^2 \times 0{\cdot}4 = 0{\cdot}432$
„ „ „ 1 „ „ „ $= 3pq^2$
$= 3 \times 0{\cdot}6 \times 0{\cdot}4^2 = 0{\cdot}288$
„ „ „ 0 „ „ „ $= q^3 = 0{\cdot}4^3 = 0{\cdot}064$

Total probability $= 1{\cdot}000$

So the possibility of the visited farms reflecting the overall balance of 60% with dairying, i.e. approximately 2 farms out of 3 with dairying, is less than 50%, for there is a more than 20% probability that all the 3 farms will include dairying, and more than a 30% chance that no more than one of the 3 farms will include dairying. Figures such as these are a salutary warning against basing general conclusions on a too limited study and this whole theme of the size of the sample for study and the degree of accuracy that this provides must be taken up at greater length in Chapter 6. The above example will then be considered again in more detail.

Probability and the Poisson Frequency Distribution

In all these examples of assessing the probability with which given conditions occur in a specific number of occurrences the data have always been such that they could be divided into those occurrences when the given conditions *did* occur and those when they did not. Probability values on an overall basis could thus be ascribed to both sets of conditions, under the terms p and q. In some cases, however, data are concerned with isolated events in time when although it is possible to specify the number of times certain conditions *did* occur it is *not* possible or not sensible to say how often they did *not*. For example, it is possible to consider the number of times that hail falls or fog occurs in a month, or the number of times that a river floods in a winter or a wet season, but seeking to know how many times these events did *not* occur is neither sensible nor feasible.

In such studies as these the data are always discrete, i.e. whole numbers, the frequency distribution is usually skew and there is a limit to the possibilities in one direction because of zero values and perhaps in the other because of magnitude. The question that normally requires solution here is the probability with which different numbers of these occurrences are likely to occur. Thus it may be desired to know the probability of a particular river flooding 0, 1, 2, 3, 4 or 5 times in a wet season. Here the limiting factor of zero values is clearly important, while again it is unlikely that the values could continue increasing indefinitely. It would, of course, be possible to assess these probabilities by calculating the average and standard deviation values, obtaining overall probabilities from the normal distribution function and then calculating probabilities from the binomial distribution, as has been done with the examples worked out above. With a set of data which is markedly skew, however, the probabilities from the normal distribution function would be of only generalized reliability, so that transformation would be necessary; therefore a probability distribution which closely approximates to a skew distribution would provide a better estimate of probability.

For example, consider the sample data set out below concerning the number of times a river floods in a wet season. Clearly the frequency will differ from one year to another, and the number of years in which 0, 1, 2, 3, 4 or 5 floods occurred during a period of 100 years is given in the following table.

No. of years	No. of floods
24	0
35	1
24	2
12	3
4	4
1	5

With the total number of 140 floods during the 100 years, the average number of floods per year is 1·4. Calculation will also show that the standard deviation for these data is 1·15. By applying the formula $d = \frac{x - \bar{x}}{\sigma}$ and referring to the table of the normal distribution function (Table XII) for given values of x, it is found that the estimated probabilities by this method greatly overestimate the frequency of years with many floods and underestimate the frequency of years with few floods when compared to the sample data themselves. Any estimate by the binomial distribution using these values for p and q will therefore equally be too divorced from reality to be of real value.

To be able to postulate probability values in such a case it is necessary to use a third technique, this being based on the Poisson distribution. This distribution utilizes the mathematical constant that is written as e; its value is the limit approached by $\left(1 + \frac{1}{n}\right)^n$ as n becomes very large, so that, correct to four decimal places, it is 2·7183. This is used in a series of successive terms which express the probability of 0, 1, 2, 3, 4, etc. events occurring. These terms are as follows;

$$e^{-z}; \quad z.e^{-z}; \quad \frac{z^2}{2!}.e^{-z}; \quad \frac{z^3}{3!}.e^{-z}; \quad \frac{z^4}{4!}.e^{-z}$$

In these terms
e is the value 2·7183 indicated above;
z is the average value for the set of data;
e^{-z} is the same as writing $\frac{1}{e^z}$
! indicates that it is the 'factorial' of the number concerned (see p. 74).

By calculating the values for these terms it is possible to evaluate the probabilities of 0, 1, 2, etc. events occurring, without first calculating the standard deviation or making any other prior assessments. One thing is required, however, namely that the average or expected number, i.e. z, should be constant (or virtually so) from trial to trial, i.e. from one set of years to another.

This formula can be applied to the present data as follows:

$$e^{-z} = 2{\cdot}7183^{-1{\cdot}4} = \frac{1}{2{\cdot}7183^{1{\cdot}4}} = 0{\cdot}2466 = \text{probability of 0 floods}$$

$$z.e^{-z} = 1{\cdot}4 \times 0{\cdot}2466 = 0{\cdot}3452 = \text{probability of 1 flood}$$

$$\frac{z^2}{2!}.e^{-z} = \frac{1{\cdot}4^2}{2} \times 0{\cdot}2466 = 0{\cdot}98 \times 0{\cdot}2466 = 0{\cdot}2417 = \text{probability of 2 floods}$$

$$\frac{z^3}{3!}.e^{-z} = \frac{1{\cdot}4^3}{6} \times 0{\cdot}2466 = 0{\cdot}458 \times 0{\cdot}2466 = 0{\cdot}1127 = \text{probability of 3 floods}$$

$$\frac{z^4}{4!}.e^{-z} = \frac{1{\cdot}4^4}{24} \times 0{\cdot}2466 = 0{\cdot}1602 \times 0{\cdot}2466 = 0{\cdot}0395 = \text{probability of 4 floods}$$

$$\frac{z^5}{5!}.e^{-z} = \frac{1{\cdot}4^5}{120} \times 0{\cdot}2466 = 0{\cdot}0449 \times 0{\cdot}2466 = 0{\cdot}0110 = \text{probability of 5 floods}$$

$$0{\cdot}9967 = \text{approximate total probability}$$

To indicate the extent to which this method does provide a valid index of the probability with which these events occur, the events themselves, the probability values, the frequency which this implies over a hundred years, and the actual sample values presented earlier are all tabulated below.

Number of floods per wet season	0	1	2	3	4	5
Probability value	0·2466	0·3452	0·2417	0·1127	0·0395	0·0110
Probable frequency per hundred years	25	35	24	11	4	1
Actual frequency in the specified hundred years	24	35	24	12	4	1

Thus a *very* close approximation to the actual conditions was provided by the Poisson distribution when applied to this sort of data. It may also have been observed that, the standard deviation being 1·15, the variance was therefore 1·32. This variance is almost the same as the mean value of 1·4, and this coincidence of the average and the variance is the hall-mark of data which fit the Poisson distribution.

Apart from such a study of isolated events in time it is also possible to analyse in this way isolated events in space or distance. For example, it may be desirable, when studying the impact of transport facilities on industrial location, to consider the relative frequency with which industrial premises occur in close proximity to railway stations. This could then perhaps be compared to the frequency with which such premises occur near port facilities and trunk road junctions. Such problems of comparison will be considered later in Chapters 8–10. In the case of the railway stations, a count could be made to discover how many industrial premises occur near each of a series of sample stations; the method of choosing the stations to be studied will be outlined in Chapter 7. For now, assume that the following figures were obtained.

No. of stations	No. of industrial premises near that station
182	0
91	1
23	2
3	3
1	4

In this example there are 300 stations and the average number of industrial premises per station is 0·5. Further calculation will show that the variance of this set of data is 0·503. As this is virtually the same as the mean it is therefore possible to use the Poisson distribution to make an assessment of the probability with which given numbers of premises will occur near each station. The normal curve and the binomial distribution could not be used in this case, for with a standard deviation of 0·71 any probabilities obtained in that way would underestimate the occurrence of few premises and overestimate the frequency of many premises. The Poisson distribution values can be obtained as follows:

Term	Value	Probability	No. of occurrences Calculated	Observed
e^{-z}	$= 2{\cdot}7183^{-0{\cdot}5}$	$= 0{\cdot}6065$	$=$ 181·98	182
$z.e^{-z}$	$= 0{\cdot}5 \times 2{\cdot}7183^{-0{\cdot}5}$	$= 0{\cdot}3032$	$=$ 90·97	91
$\frac{z^2}{2!}.e^{-z}$	$= \frac{0{\cdot}5^2}{2} \times 2{\cdot}7183^{-0{\cdot}5}$	$= 0{\cdot}0758$	$=$ 22·75	23

$$\frac{z^3}{3!}.e^{-z} = \frac{0{\cdot}5^3}{6} \times 2{\cdot}7183^{-0{\cdot}5} = 0{\cdot}0126 \qquad = 3{\cdot}79 \qquad 3$$

$$\frac{z^4}{4!}.e^{-z} = \frac{0{\cdot}5^4}{24} \times 2{\cdot}7183^{-0{\cdot}5} = 0{\cdot}0016 \qquad = 0{\cdot}47 \qquad 1$$

The close relationship of these estimated values to the sample ones is clear. In most cases, of course, the relationship is nowhere near as marked, the variance not being as close to the mean as it was in this case. When the variance differs too greatly from the mean it is still possible to use the Poisson distribution by means of an adjustment to the formula, but this can be followed up by the reader in more advanced texts if he so requires.

For cases similar to those presented here, however, the labour of calculating the various probability values by the formula given can be eliminated by the use of the appropriate probability paper. Thus in the flood example outlined on p. 80, it was seen that the *average* flood frequency ($\underset{\sim}{z}$) was 1·4 floods per annum. This value can then be referred to the base line of Fig. 23, and on projection upwards it intersects the lines showing the number of occurrences (listed to the right of the graph). Against each such intersection, a probability value (to the left of the graph) can be read off. For example, with an average of 1·4 and a frequency of occurrence of 1, the appropriate probability value is approximately 0·75. This implies that there is a 75% probability of *1 or more* occurrences, and therefore a 25% probability of less than 1 occurrence, i.e. of no occurrences. Reference to the results given on p. 81 will show that this is the same answer as that obtained by calculation. A more precise comparison between calculated and graphical values is set out below:

From Graph (Fig. 23) Occurrences	Probability	No. of occurrences	Probability from graph (Fig. 23)	Probability calculated (p. 81)
1 or more	0·753	0	1–0·753 = 0·247	0·247
2 or more	0·408	1	0·753–0·408 = 0·345	0·345
3 or more	0·165	2	0·408–0·165 = 0·243	0·242
4 or more	0·054	3	0·165–0·054 = 0·111	0·113
5 or more	0·014	4	0·054–0·014 = 0·040	0·040
6 or more	0·003	5	0·014–0·003 = 0·011	0·011

Thus, despite the slightly different form in which the initial results from the probability paper are obtained, it is a simple procedure by which to obtain the results previously arrived at by laborious calculation. The example on p. 82 concerning the frequency of

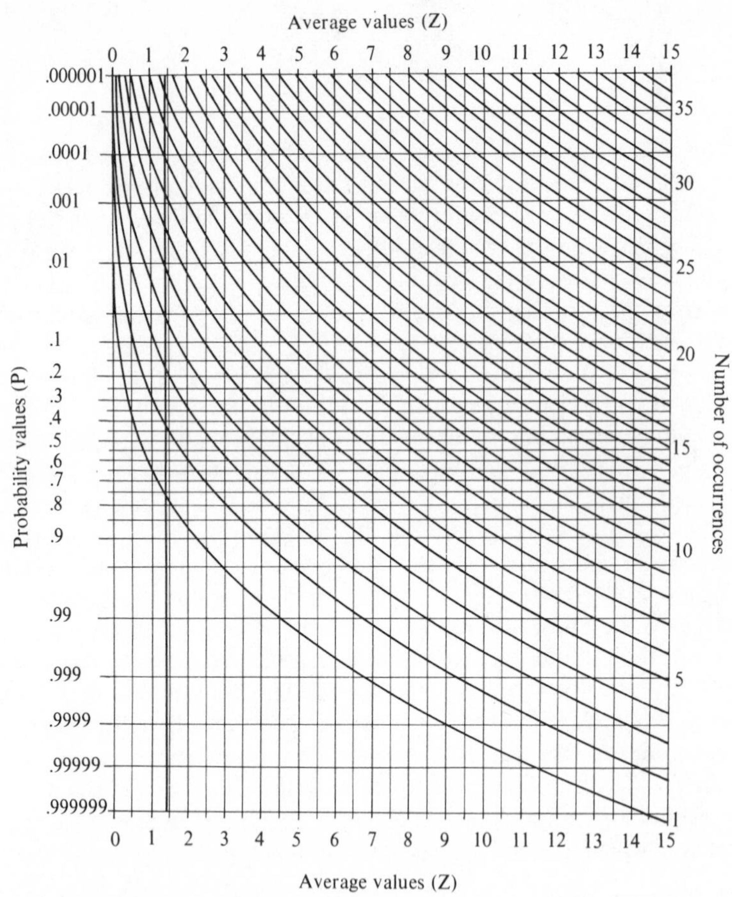

Figure 23. Poisson probability paper

industrial premises near railway stations can also be worked out in the same way, and the reader is advised to carry out such a check by means of the probability paper in Fig. 23.

Throughout this and the previous chapter the average, standard deviation and variance values, the methods of calculating which were outlined in Chapters 2 and 3, have been put to some practical use beyond the simple representation of the basic parameters of a set of data. Especially they have been employed in the assessment of the probability with which given conditions may be expected to occur. In order to do this it has been shown to be necessary to allocate the data to one of several frequency distributions, the one chosen being partly conditioned by the character of the data and partly by the problem that it is desired to solve. In all cases, however, the aim of assessing probabilities has been to obtain from a limited set of sample data information of what is likely to occur throughout a much larger total population. As it is so useful to be able to obtain an assessment about conditions in a large body of data by analysing a relatively small body, and also as it is often the case that only a 'sample' of conditions is in fact available, it is therefore essential that the characteristics and limitations of working on sample data be understood. That is the purpose of Chapters 6 and 7.

Before leaving the present chapter, however, there is one general point that must be stressed. In the three methods considered above, the focus has been upon the simple theme of assessing from information available from sample data, the probability of given events occurring in the total population. This has direct practical relevance itself, and is also linked to the important theme of hypothesis testing in Chapters 8–10. At research level, however, beyond the scope of this present book, probability theory has been used increasingly as the means of generating locational models and theories based on probabilistic rather than deterministic assumptions. In this context, both the binomial and the Poisson distributions (plus their more complex derivatives) are of fundamental importance. The results of such work are illustrated in many of the publications listed under *probability studies and models* in the bibliography.

CHAPTER 6

CHARACTERISTICS OF SAMPLES

Sample and Population Parameters

As indicated previously, in most of the methods so far outlined there is the implied assumption that the values obtained, especially in terms of mean and deviation, apply to an infinitely long series of data. This long series of data is referred to as the *population*, and the parameters obtained are thus, for example, the *population mean* and the *population standard deviation.* More concisely, at times these may be called the *true* mean, etc., this term thus implying that it is the value which would be obtained from analysing the whole body of data concerning the phenomenon under study. The values that in practice are obtained are usually based on only part of the body of data, this being the result either of a deliberate choice or because no more data are available, i.e. these values are based on only a *sample* of the conditions. Thus what is usually obtained is not the true or population mean but the *sample mean*; the same applies to the standard deviation too. Before proceeding to any assessment of the differences between different series of data, or to any further conclusions based on the mean and the standard deviation, it is therefore essential that some thought be given to the relationship between these sample statistics and the true parameters.

The relationship that may be expected to hold true between sample and population parameters is partly conditioned by the size of the sample and partly by the method of obtaining the sample. Ideally the choice of sample would be purely random, i.e. without any bias whatsoever, and simply as a chance selection of so many items out of the body of data. The means by which a random choice may be made will be outlined in Chapter 7; suffice it to say at this stage that such a sample should give as true and representative a cross-section of the population as is permitted by the size of the sample. In many cases, however, especially in geographical analyses, such a random selection is either not possible or not desirable for other reasons. The general concepts on which sampling techniques are based are nevertheless best explained in terms of random sampling, and the modifications necessitated by non-random samples can then be presented afterwards.

Given that the sample is a random one, the major factor controlling the relationship between sample and population values is thus the size of the sample. The influence of this can probably best be seen if a slight digression be made to consider again the frequency distribution curve of a normal distribution. In Fig. 24 the curve for the individual items of a set of data (i.e. when $n = 1$) is the lowest and most broadly based of those which are shown, while the average obtained from these individual items is shown as being centrally placed. Suppose, however, that instead of considering *individual* items, these data were first grouped arbitrarily into groups of 3 items each, i.e. that *samples of* 3 *items each* were obtained by random sampling, and that the average were to be obtained for each of these samples of 3 items. It would then be possible to plot a distribution curve for these 'means of 3 items' and an overall average value obtained. This average would be the same as that for the individual items, but the curve would differ. When taking the samples it would be unlikely that in all cases all three items in the sample would lie on the same side of the average. With the averaging of these 3 items the likely range of values of 'means of 3 items' would be less than that for individual items, so that the values would cluster more closely around the average. So although the average of the '3 item sample' data would be the same as that of the individual items, its variance and standard deviation would be less. This is shown diagrammatically in Fig. 24 by the second lowest of the curves. This lesser degree of scatter of sample means than of individual values around the average applies no matter what size of the sample is taken. On the other hand, the greater the number of items in the sample means, the smaller will be the scatter of these sample means, as is shown in Fig. 24 by the topmost curve for '10 item samples'. This indicates that the variance of these distribution curves based on sample means is related to the

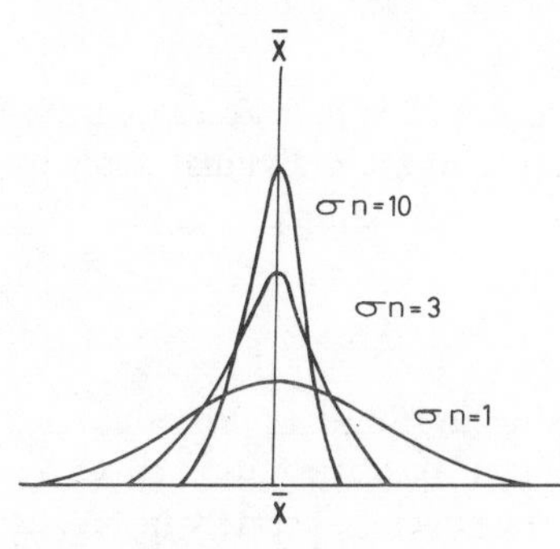

Figure 24. Distribution curves of sample means of n items

number of items in the sample. This relationship is expressed as follows:

$$\frac{\text{variance of sample means with } n \text{ items per sample}}{} = \frac{\text{variance of individual items}}{\text{number of items per sample}}$$

or more briefly: $\text{var.}_n = \dfrac{\sigma^2}{n}$

Furthermore, as the standard deviation is the square root of the variance the standard deviation of sample means with n items per sample can be obtained as follows:

$$\sigma_n = \sqrt{\text{var.}_n} = \sqrt{\frac{\sigma^2}{n}} = \frac{\sigma}{\sqrt{n}}$$

i.e. the standard deviation of a distribution of sample averages is obtained by dividing the standard deviation of the individual items by the square root of the number of items in the sample.

Sampling or Standard Error

The greatest value of this relationship to sampling procedure lies in a corollary from the above argument. If the distribution curve for the 'means of samples of 10 items' is considered separately in Fig. 24 it is seen that it is a *normal* curve symmetrical about an average value which is the same as the average value for the overall data given by the individual items. It can therefore be argued that, because of the characteristics of the normal distribution, it is *extremely improbable* that any one 'mean of a sample of 10 items' will differ from this overall average by more than 3 standard deviations, i.e. by more than $3\left(\frac{\sigma}{\sqrt{10}}\right)$, and that it is *unlikely* that it will differ from this overall average by more than 2 standard deviations, i.e. by more than $2\left(\frac{\sigma}{\sqrt{10}}\right)$. If this is so, the reverse argument can also be applied, namely that if any given 'mean of a sample of 10 items' is known then the overall or true mean is *unlikely* to differ from this sample mean by more than $2\left(\frac{\sigma}{\sqrt{10}}\right)$ and it is *extremely improbable* that it will differ from this sample mean by more than $3\left(\frac{\sigma}{\sqrt{10}}\right)$. Thus, if a sample mean is obtained it is possible to indicate the limits within which the true mean must lie with a given percentage probability, i.e.

the true mean $\bar{X}$ = the sample mean $\bar{x} +/- 2\left(\frac{\sigma}{\sqrt{n}}\right)$ with a 95·45% probability;

or = the sample mean $\bar{x} +/- 3\left(\frac{\sigma}{\sqrt{n}}\right)$ with a 99·7% probability.

In most cases the true mean will lie closer to the sample mean than these values, for these only indicate the limits beyond which it is unlikely that the true mean will lie.

An example of this sort of application will help to stress its implications. Suppose that a study is being made of farming over a large area and an assessment is required of the average size of farm holdings. The total number of farms is so large that it is decided to study only a sample of these farms. Provided that this sample is truly random, picked in a way that will be outlined in Chapter 7, it would be possible to assess the limits within which the true mean should fall with a known percentage probability. The accuracy of this or of any sample is, as indicated above, related to the size of the sample, and thus *not* to the percentage of the total data which is included in the sample. Given that the variance of the sample mean is expressed by $\frac{\sigma^2}{n}$ it is clearly the magnitude of n which is important, whether this be 90% or 9% of the total of occurrences. In the present example it could be that a sample of 200 farms is to be taken. From these it is found that the sample average acreage is 90 acres and that the sample standard deviation is 7 acres. The calculation of the limits of the true mean is thus as follows:

no. of items (n) = 200 sample mean ($\bar{x}$) = 90

sample standard deviation (indicated by s rather than σ) = 7

true mean = $\bar{X}$

Thus, $\bar{X} = \bar{x} +/- \frac{s}{\sqrt{n}}$ with a confidence limit, i.e. a percentage probability of being correct, of c. 68%

$$= 90 +/- \frac{7}{\sqrt{200}} = 90 +/- 0{\cdot}5 \text{ (actually } 0{\cdot}496\text{)}$$

i.e. $\bar{X}$ lies between 89·5 and 90·5 with a c. 68% probability.

Again, $\bar{X} = \bar{x} +/- 2.\frac{s}{\sqrt{n}}$ with a confidence limit of c. 95%

$$= 90 +/- (2 \times 0{\cdot}5) = 90 +/- 1{\cdot}0$$

i.e. $\bar{X}$ lies between 89·0 and 91·0 with a c. 95% probability.

Further, $\bar{X} = \bar{x} +/- 3.\frac{s}{\sqrt{n}}$ with a confidence limit of 99·7%

$$= 90 +/- (3 \times 0{\cdot}5) = 90 +/- 1{\cdot}5$$

i.e. $\bar{X}$ lies between 88·5 and 91·5 with a 99·7% probability.

Thus it can be seen that limits can be set to the true mean value, and that these limits are wider the more stringent are the probability values. This value which controls these limits, i.e. $\frac{s}{\sqrt{n}}$, is known in this connection as the *standard error of the mean.*

Although this does provide an estimate of the limits of the true mean, it equally stresses the limitations implicit in a sample mean even when it is based on a sample as large as 200. If a sample ten times as large were taken, i.e. if $n = 2{,}000$, it would be found that the standard error of the mean (S.E. $\bar{x}$) equals 0·157 acres instead of the value of 0·5 acres based on 200 items. Thus by a sample ten times as large the 'error' is reduced to about a third of its size, and the limits of the true mean could then be set as being between 89·53 and 90·47 with a 99·7% probability. It can here be seen that to alter the probability limits for these values from approximately 68% to 99·7% requires a tenfold increase in the size of the sample (and the work associated with it). Much of the art of sampling lies in choosing a size of sample that will give an answer with the desired degree of accuracy and probability with the minimum sample size. However, if a certain degree of accuracy is required it must necessarily mean a certain-sized sample—there is no satisfactory way of getting an adequate answer with an inadequately sized sample.

A comparable sort of standard error can also be obtained for the standard deviation. This *standard error of the standard deviation* is obtained by the expression $\frac{s}{\sqrt{2n}}$, from which the degree of uncertainty inherent in the estimate of the standard deviation from a sample can be obtained. So in the farm acreage example above the true standard deviation can be assumed to lie within the following limits with the following degrees of probability:

true standard deviation σ = sample standard deviation $s +/- \frac{s}{\sqrt{2n}}$

with a 68% probability

$$\text{i.e. } \sigma = 7 +/- \frac{7}{\sqrt{2 \times 200}} = 7 +/- 0{\cdot}35$$

i.e. the true standard deviation lies between 6·65 and 7·35 with a 68% probability.

Similarly it would be found that the true standard deviation lies between 6·3 and 7·7 with a c. 95% probability and between 5·95 and 8·05 with a 99·7% probability. Again, if the sample were to be increased to 2,000 items then the true standard deviation would be seen to lie between 6·67 and 7·33 with a 99·7% probability. The accuracy of these statements can be readily checked by the reader by calculating the standard error of the standard deviation on the basis of 2,000 items, a sample mean of 90 and a sample standard deviation of 7.

Best Estimates, Small Samples and Small Populations

In all these calculations of standard errors which have so far been presented one assumption has been made which is not strictly justified. From the diagram in Fig. 24, the mean value and the standard deviation value which led to the expression that $\sigma_n = \frac{\sigma}{\sqrt{n}}$ (p. 88) were the mean and standard deviation of the *total* population. In the above samples, however, it is the mean and standard deviation of only the *one* sample which is used. This is often done through sheer necessity for only the sample data may be available. Nevertheless, in order to be able to apply the method of obtaining the standard error with some justification, an *estimate* should be made of the *true* standard deviation. This process is usually referred to as making a *best estimate*, and it is done by applying a correction to the sample standard deviation. This correction, which is sometimes called *Bessel's correction*, is $\sqrt{\frac{n}{n-1}}$ for changing the sample standard deviation to the best estimate of the standard deviation, and it is $\frac{n}{n-1}$ for changing the sample variance to the best estimate of the variance. There are thus three possible values for which the term standard deviation is used, and each has its own symbol. There is the *sample* standard deviation (s), the *true* or *population* standard deviation (σ), and the *best estimate* of the standard deviation ($\hat{\sigma}$)—such a circumflex over a symbol always indicates a 'best estimate'.

It is possible to apply this correction to the values used in the previous example. The sample standard deviation in that case was 7. This must therefore be multiplied by $\sqrt{\frac{n}{n-1}}$

$$\hat{\sigma} = s \, . \sqrt{\frac{n}{n-1}}$$

$$= 7 \times \sqrt{\frac{200}{200-1}} = 7 \times 1{\cdot}0025$$

$$= 7{\cdot}0175$$

This best estimate of 7·0175 can therefore be inserted in the calculation of the standard error of the mean which becomes

$$\text{S.E.}\ \bar{x} = \frac{\hat{\sigma}}{\sqrt{n}} = \frac{7{\cdot}0175}{\sqrt{200}} = 0{\cdot}498$$

The difference between this and the value of 0·496 on p. 90 is negligible, because of the size of the sample. It is clear that the larger the sample the closer will the correction $\sqrt{\frac{n}{n-1}}$ approximate to unity, while if the sample is small the value of $\sqrt{\frac{n}{n-1}}$ will be considerably above unity and will therefore markedly affect the size of the standard error. This is but one of the problems associated with small samples, which will be examined further in later pages.

This extra calculation of the best estimate of the standard deviation can in fact be avoided by integrating the correction factor $\sqrt{\dfrac{n}{n-1}}$ into the standard deviation formula. The correctness of this integration can more clearly be seen if it is first effected for the variance rather than the standard deviation. So, if the sample variance is s^2 the conversion to the best estimate of the variance ($\hat{\sigma}^2$) is made as follows:

$$\hat{\sigma}^2 = s^2 \times \frac{n}{n-1}$$

$$= \frac{\Sigma (x - \bar{x})^2}{n} \times \frac{n}{n-1} = \frac{\Sigma (x - \bar{x})^2}{n-1}$$

As the standard deviation is the square root of the variance it follows that the best estimate of the standard deviation may be obtained from a sample by direct calculation from the formula

$$\hat{\sigma} = \sqrt{\frac{\Sigma (x - \bar{x})^2}{n-1}}$$

Thus the calculation in the above example would be

$$\hat{\sigma} = \sqrt{\frac{9{,}800}{199}} = \sqrt{49{\cdot}246} = 7{\cdot}0175$$

This gives the same answer as by the application of the correction after calculating the sample standard deviation (p. 92). As this difference between the sample and best estimate values may well be of significance at times, it is always essential, *when using a set of data as a sample of a larger body of data*, to insert $n - 1$ for n in the standard deviation calculation whichever of the formulae on pp. 27-31 is used, i.e.

$$\hat{\sigma} = \sqrt{\frac{\Sigma(x-\bar{x})^2}{n-1}} = \sqrt{\frac{n}{n-1}\left(\frac{\Sigma x^2}{n} - \bar{x}^2\right)} = \sqrt{\frac{1}{n-1}\left(\Sigma x^2 - \frac{(\Sigma x)^2}{n}\right)}$$

Although the application of this correction helps to counterbalance any underestimate of conditions introduced by a sample which is not very large, there is the need *when samples are really small* for a further modification to be made, this time to the actual use of the standard error. In small samples it is no longer safe or justified to assume that, for example, values will lie within two standard deviations of the mean with a 95% probability. In other words, the probability values of the normal curve cannot be assumed to apply to the sample even though the full body of the data fits the normal curve. Instead use should be made of *Student's t* distribution. This will be considered more fully in Chapter 8. For now it is sufficient to refer to the graph in Fig. 30 (p. 144). For this it is necessary first to obtain the value $(n - 1)$ which is here known as the 'degrees of freedom' (see p. 142), and then to read off against this the 't' value for the required probability level. Thus if on the normal curve a 95% probability of values lying within the two standard deviation limits would be used, then 't' is read off at the 5% level on Fig. 30. The value thus obtained, which will be somewhat larger than 2, is then used in the true mean calculations instead of the value of 2 itself, when multiplying the standard error value. So, whereas with a large sample the limits of the *true mean* ($\bar{X}$), defined with a 95% probability, would be obtained from

$$\bar{X} = \bar{x} +/- 2.\frac{\hat{\sigma}}{\sqrt{n}}$$

with a small sample the formula would become

$$\bar{X} = \bar{x} +/- t.\frac{\hat{\sigma}}{\sqrt{n}}$$

The same would be true when assessing the *true standard deviation*, which with small samples would thus be

$$\sigma = s +/- t.\frac{\hat{\sigma}}{\sqrt{2n}}$$

The differences which these modifications of the formula may introduce into assessments of the limits of the true mean and standard deviation can most readily be appreciated if one set of sample data is analysed by the several methods outlined above and the resulting assessments are then compared. As a practical example, assume that a study is being made of the number of people in a series of parishes or communes over a large area. The total number of units is considerable, but some reasonable degree of similiarity in population size, etc. can be expected on the basis of prior knowledge of the area. It is therefore decided to make a rapid sample analysis of values before making a full study, so that any obvious problems can be foreseen. For this purpose a sample is chosen at random (see p. 104), totalling as few as only 10 communes. From this sample the following statistics are calculated:

number of items (n) = 10 communes

sample average ($\bar{x}$) = 350 people per commune
sample standard deviation (s) = 15 people

From these it would be possible to calculate the limits of the true mean with a 95% probability of being right by the formula

$$\bar{X} \text{ (with a 95\% probability)} = \bar{x} +/- 2.\frac{s}{\sqrt{n}} = 350 +/- \frac{2 \times 15}{\sqrt{10}}$$

$$= 350 +/- \frac{30}{3{\cdot}16} = 350 +/- 9{\cdot}5 = \underline{340{\cdot}5 \text{ to } 359{\cdot}5}$$

This, however, fails to take into account the fact that only the sample standard deviation is being used and that the best estimate of this parameter should in fact be employed. This is therefore obtained as follows:

best estimate of standard deviation ($\hat{\sigma}$)

$$= s.\sqrt{\frac{n}{n-1}} = 15 \times \sqrt{\frac{10}{9}} = 15 \times \sqrt{1{\cdot}11}$$

$$= 15 \times 1{\cdot}055 = 15{\cdot}825$$

With $\hat{\sigma}$ inserted for s the assessment of the limits of the true mean becomes

$$\bar{X} \text{(with a 95\% probability)} = \bar{x} +/- 2.\frac{\hat{\sigma}}{\sqrt{n}} = 350 +/- \frac{2 \times 15{\cdot}825}{3{\cdot}16}$$

$$= 350 +/- 10 = \underline{340 \text{ to } 360}$$

Such an assessment is strictly only applicable if the sample is a large one, but in this case it is small (this term frequently being taken to imply 10 items or less, although the methods are often applied to rather larger samples too, to be on the safe side). Therefore the frequency values of the normal distribution should be replaced by those of the Student's t distribution. Referring to Fig. 30, it is first necessary to obtain what are called the 'degrees of freedom', i.e. $n - 1$, which in this case is $10 - 1 = 9$. The value for t for 9 degrees of freedom is then read off at the 5% level, this being virtually the equivalent of the 2 standard deviation probability on the normal curve. This gives a value for t of 2·4 at this 5% level. It is this value which must now replace the 2 in the formula. Thus, bearing in mind the fact that this is only a small sample, plus the need to correct in terms of the best estimate of the standard deviation, the limits of the true mean, with a 95% probability of being right, are:

$$\bar{X}\text{(with a 95\% probability)} = \bar{x} +/- t.\frac{\hat{\sigma}}{\sqrt{n}} = 350 +/- \frac{2{\cdot}4 \times 15{\cdot}825}{3{\cdot}16}$$

$$= 350 +/- 12 = \underline{\underline{338 \text{ to } 362}}$$

In this way it can be seen that the limits of the true mean are in fact wider than might be assumed at first, and it is the latter set of values which should be used. In terms of the present example it means that by considering only ten communes, and assuming that these are representative of the whole data, the overall average (i.e. true mean) population per commune or parish can be assessed within reasonable limits, i.e. it will almost certainly lie between 338 and 362 persons. Such an assessment can well provide a useful indication of the order of magnitude within which working will take place, and also of the order of detail that may be required to enable significant differences to be appreciated. Furthermore, this example also indicates that the closeness with which sample values will approximate to true values is controlled by several variables. The difference between sample and true values will increase as the stringency of the percentage probability of being right is increased, as the best estimate of the standard deviation increases and as the size of the sample decreases. As the second of these variables is inherent in the body of the data, it is only in the first and the last that there is some element of conscious choice. This choice is exercised in terms of the character of the analysis, its purpose, and the degree of accuracy required.

Before considering how a decision can best be made concerning the most suitable size of sample, one further theme must be outlined. Suppose that when sampling it was found that the size of the *total population* was very small, in contrast to the earlier examples where it was the sample size that was small. In such a case the best estimate of the standard deviation would be virtually the true value, i.e. $\hat{\sigma} = \sigma$. Therefore the standard error would be less than the usual formula would suggest, and the standard error calculated in the normal way must be modified by a factor related to the proportion of the population forming the sample. This proportion is known as the 'sampling fraction'. The factor used is $\sqrt{1 - f}$ where f is the sampling fraction. This means that if *all* the population were to be included in the sample, then the sampling fraction f would be unity, the correction factor would be 0 and therefore the standard error would also be 0.

With this factor included, the standard error of the mean for a random sample of a small total population is

$$\text{S.E. } \bar{x} = \frac{\hat{\sigma}}{\sqrt{n}} \cdot \sqrt{1-f} \quad \text{or} \quad \sqrt{\frac{\hat{\sigma}^2}{n}} \cdot \sqrt{1-f} \quad \text{or} \quad \sqrt{\frac{\hat{\sigma}^2}{n} \cdot (1-f)}$$

So if it were found from a small total population that $\hat{\sigma} = 40$, that the number of items in the sample $n = 4$ and that the sampling fraction $f = \frac{1}{5}$, the standard error would *not* be

$$\frac{\hat{\sigma}}{\sqrt{n}} = \frac{40}{\sqrt{4}} = \frac{40}{2} = 20$$

$$\text{but } \frac{\hat{\sigma}}{\sqrt{n}} \cdot \sqrt{1-f} = 20 \times \sqrt{1-0{\cdot}2} = 20 \times \sqrt{0{\cdot}8} = 20 \times 0{\cdot}9 = 18{\cdot}0$$

(approximately).

In this way the standard error is reduced for the same size of sample, but *only* if the total population is itself not large.

Specification of Sample Size

It has been indicated earlier that it is often of very great value to be able to judge the minimum size of sample that will ensure that the true mean is obtained to within given limits. For example, in the case outlined on pp. 94–96 it is considered desirable to establish the true mean's limits with a probability of 95%. Also, from a small sample such as ten items the best estimate of the standard deviation has been calculated as 15·825. The range within which the true mean lies is too wide if only ten items are included in the sample, and it is decided that to be able to make any useful general judgments from a sample the true mean needs to be defined to within +/−5 of the sample mean (at the 95% probability level). The question therefore is what size of sample needs to be taken to give this degree of accuracy under these conditions; i.e. assuming on the evidence of the 10 item sample that for the required degree of accuracy the sample will *not* be a small one, what size of sample will yield a standard error of 2·5? The formula for the standard error is

$$\text{S.E.} = \frac{\hat{\sigma}}{\sqrt{n}}$$

and this must equal a desired value (d). So

$$\frac{\hat{\sigma}}{\sqrt{n}} = d$$

$$\frac{1}{\sqrt{n}} = \frac{d}{\hat{\sigma}}$$

$$\sqrt{n} = \frac{\hat{\sigma}}{d}$$

$$n = \left(\frac{\hat{\sigma}}{d}\right)^2$$

In the present example, when $\hat{\sigma} = 15{\cdot}825$ and $d = 2{\cdot}5$,

$$n = \left(\frac{15{\cdot}825}{2{\cdot}5}\right)^2 = 6{\cdot}33^2 = 40 \text{ items in the sample.}$$

As a check to show that a sample of that size would give the desired result, provided that the characteristics of the new larger sample were the same as those of the previous smaller one, the following calculation can be made in terms of $n = 40$.

$$\bar{X} \text{ (at 95\% probability)} = \bar{x} +/- 2.\frac{\hat{\sigma}}{\sqrt{n}}$$

$$= 350 +/- \frac{2 \times 15{\cdot}825}{\sqrt{40}}$$

$$= 350 +/- \frac{31{\cdot}65}{6{\cdot}33} = 350 +/- 5$$

$$= 345 \text{ to } 355 \text{ persons}$$

This formula for calculating the size of the sample required for given conditions can always be applied to data based on random sampling, when the population is virtually normal in distribution and when some best estimate of the standard deviation has been made. It must be remembered, however, that it will still provide only an *approximate* answer to the desired sample size, for it is extremely unlikely that the same average and standard deviation values will apply to both samples.

Standard Errors and Probability Assessments

This theme of standard errors resultant on the use of sample data is also of considerable relevance in connection with the probability assessments from the normal curve discussed in Chapter 5, for the average and standard deviation for one period or body of data are there being used as the basis for assessing the probability of specified conditions occurring in a longer period or set of data. In this way, the average and standard deviation values from which the assessments are made are in effect derived from sample data, and they therefore incorporate a sampling or standard error.

For example, if one analyses the 1936–55 annual rainfall record for Lourenço Marques, the following values are obtained:

$\bar{x} = 815$ mm $\quad \hat{\sigma} = 225$ mm $\quad n = 20$

From these, it is possible to estimate, by the methods given in Chapter 5,

(*a*) The probability of obtaining less than 1,000 mm. in a year, for which

$d = 0{\cdot}82 =$ probability 79·4%.

(*b*) The annual rainfall likely to be exceeded 80% of the time, which is 626 mm.

The standard errors of these two estimates are compounded of the standard errors of both the sample mean and the sample standard deviation. In the case of (*b*) above, this can be obtained as follows:

$$\text{S.E. of critical value} = \sqrt{\begin{matrix}\text{variance of}\\ \text{sample mean}\end{matrix} + \begin{matrix}\text{variance of sample}\\ \text{standard deviation}\end{matrix}}$$

$$= \sqrt{\frac{\hat{\sigma}^2}{n} + \frac{\hat{\sigma}^2}{2n}} = \sqrt{\frac{3\hat{\sigma}^2}{2n}}$$

$$= \sqrt{\frac{3 \times 225^2}{2 \times 20}} = \sqrt{\frac{151{,}875}{40}}$$

$$= \sqrt{3{,}797} = 61{\cdot}6 \text{ approx.}$$

As a result, the true value that will be exceeded 80% of the time can only be quoted as being between 503 mm. and 749 mm., at the 95% probability level, although 626 mm. is the best estimate.

Again, the estimate of the probability of obtaining less than

1,000 mm. is also subject to sampling error, but in this case the error is in terms of the d value from which the probability is estimated. This means that various other components enter the formula, along with the standard errors of the sample mean and sample standard deviation. With cancellations, the appropriate formula becomes:

$$\text{S.E. of } d \text{ value} = \sqrt{\frac{1}{n} + \frac{(x - \bar{x})^2}{2n\hat{\sigma}^2}}$$

$$= \sqrt{\frac{1}{20} + \frac{(1000 - 815)^2}{2 \times 20 \times 225^2}} = \sqrt{\frac{1}{20} + \frac{185^2}{40 \times 225^2}}$$

$$= \sqrt{0{\cdot}05 + 0{\cdot}0169} = \sqrt{0{\cdot}0669}$$

$$= 0{\cdot}26 \text{ approx.}$$

Therefore, the d value and its resultant probabilities must be adjusted:

d +/− 2 standard errors = 0·30 to 1·34 = 61·8% to 90·1%

Thus, although the best estimate of the probability is 79·4%, the true value can lie within fairly wide limits.

Standard Error and Sample Size with the Binomial Frequency Distribution

Of the assumptions related to the considerations so far made in this chapter, the one that must be stressed is that of the data approximating to the normal distribution. This may often not be the case, however, when probabilities are to be calculated by the binomial distribution based on a fairly small sample. On p. 78 an example of this sort was used. In an area where the overall percentage probability of farms engaging in dairying was 60% a small sample of only 3 farms was visited. The resulting probabilities of 3, 2, 1 or 0 of these farms including dairying in their activities were set out (p. 78). The frequency distribution for this is somewhat skew, as is shown diagrammatically in Fig. 25. If a larger sample had been taken then the distribution curve would have been less skew, as is shown for a sample of 10 farms also in Fig. 25. This partial correction of skewness would have been greater still if some 30 or 40 farms had been included in the sample. In this particular case the values of p and q were 0·6 and 0·4 respectively, and in such cases an almost normal curve can be obtained with a relatively small sample. If these values were 0·1 and 0·9 instead, then a far larger sample would be needed to give a near-normal curve.

In all such binomial distributions, however, the calculations of the standard error of the sample mean, or the assessment of the size of the sample required, must be effected by slightly different methods from those outlined above. A suitable example of this can be provided by outlining a problem of assessing the proportion of an area which is under irrigation, without having to account for and study every acre. The sample data will be in the form of a certain proportion of irrigated land and a certain proportion of non-irrigated land, these two proportions together giving the total size of the sample. Thus the

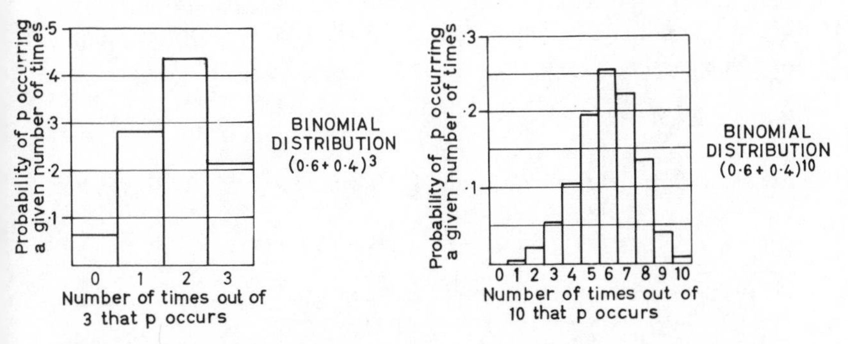

Figure 25. Effect of size of sample on the skewness of a binomial distribution

sample data are characterized by the expression $(p + q)^n$, where p is the proportion of land that is irrigated, q the proportion that is not irrigated and n the number of items in the sample. The average frequency of the given conditions, i.e. irrigated land, given by this sample may be assumed to be 30%, so that the probability is 0·3. Conversely, 70% of the sample area must therefore be non-irrigated, its probability of occurrence being 0·7. These are the p and q values in the equation. The relationship of the *true* proportions to these sample proportions, however, will depend on the size of the sample, which will affect the standard error of the sample value.

With the normal distribution this standard error is expressed as $\frac{\hat{\sigma}}{\sqrt{n}}$. This is replaced, in the case of the binomial distribution, by

$\sqrt{npq}$, which expresses the standard error in absolute terms in relation to the number of items in the sample. The values in a binomial distribution are most readily expressed, however, as a proportion or as a percentage. To obtain this the standard error given above can be multiplied by $\frac{1}{n}$ when proportions are being used and by $\frac{100}{n}$ for percentages. The resultant formulae for the standard errors are:

$\sqrt{\frac{p\,q}{n}}$ for proportions; $\sqrt{\frac{p\% \cdot q\%}{n}}$ for percentages.

In terms of the example specified earlier, the following values will obtain for the percentage standard error of the sample proportion of irrigated land. If the sample value of 30% were based on a sample of 50 items, then

$$\text{S.E. \%} = \sqrt{\frac{p\% . q\%}{n}} = \sqrt{\frac{30 \times 70}{50}} = \sqrt{\frac{2100}{50}} = \sqrt{42} = 6{\cdot}5\%$$

i.e. at the 95% level of probability, the true percentage of the whole area that is irrigated would be

$30\% +/- 2(6{\cdot}5)\% = 30 +/- 13 = 17\%$ to 43%

If, on the other hand, the sample had consisted of 300 items, then

$$\text{S.E. \%} = \sqrt{\frac{p\% . q\%}{n}} = \sqrt{\frac{30 \times 70}{300}} = \sqrt{\frac{2100}{300}} = \sqrt{7} = 2{\cdot}65\%$$

Thus, in this case the true percentage of the land that was irrigated would lie, with a 95% probability, between the following limits:

$30\% +/- 2(2{\cdot}65)\% = 30 +/- 5{\cdot}3 = 24{\cdot}7\%$ to $35{\cdot}3\%$

a more restricted range because of the larger sample.

Finally in terms of the random sampling of a binomial distribution in this way, it is often of considerable value, after an initial sample has been made, to assess the size of sample required to yield a standard error of a given magnitude. This has already been outlined for the normal distribution on p. 97 and can be calculated here as follows, with the same proviso that it gives only a general indication.

$\text{S.E. \%} = \sqrt{\frac{p\% . q\%}{n}} = d$ (where d is the desired value for the standard error)

So $\frac{p\%.q\%}{d^2} = n$

If the desired value for the standard error is set at 2%, i.e. $d = 2$, then the necessary sample size in the irrigation example is

$$n = \frac{30 \times 70}{2^2} = \frac{2100}{4} = 525 \text{ (sample size)}$$

On the other hand, if a standard error as large as 5% is adequate the sample size can be much smaller, i.e.

$$n = \frac{30 \times 70}{5^2} = \frac{2100}{25} = 84 \text{ (sample size)}$$

The size of sample required to give a standard error of 2%, and therefore an estimate of the true proportion at the 95% probability level to within +/−4%, may seem rather large at 525. This sample size, however, will apply to any size of total population, i.e. in this case to any size of area. If a study is being made of irrigated land on a large scale, 525 samples is a small price to pay for an estimate of the overall percentage value to within these close limits.

CHAPTER 7

METHODS OF SAMPLING

Methods of Random Sampling

All these considerations so far made concerning sampling have been based on the assumption that the sample itself has been a *random* sample, implying, as was stated earlier (p. 86), that the sample is an unbiased and representative cross-section of the body of data. The means by which such a random sample is obtained have not so far been considered, however. Suppose that a long list of data is available, perhaps concerning administrative units or industrial premises or climatic conditions, and it is desired to make a sample study of these data. This may be either because it is not considered worth while to analyse the full set or because a preliminary survey of this sort may enable the full study to be made more effectively. If a limited number of items were picked because they were considered 'typical', or because they stood out more clearly than the others, or because they were places known to (or near to) the person concerned, then there would be no justification for assuming that the conditions in these cases would represent the full range of conditions in the whole body of data. The sample would be what is termed 'biased', i.e. weighted in a given direction because of the way in which it was chosen. This must be very carefully guarded against, for if a choice of sample is made in this or a similar way the resulting values of mean and standard deviation conditions, and of related probability and other characteristics, will apply *only* to the sample data themselves. There will be no really adequate method of assessing the relationship between these sample characteristics and those of the population from which they were drawn, i.e. the concept of the 'standard error' which has been outlined above cannot be legitimately applied.

The choice of the sample should instead be made by reference to a table of *Random Sampling Numbers*, a short example of which, extracted from the *Cambridge Elementary Statistical Tables*, is presented in Table XIV. Thus if the data consisted of 100 items, listed in order of magnitude or in some other way, the first two columns of digits in Table XIV would be used with the numbers 00 representing

Table XIV

Random Sampling Numbers

20 17	42 28	23 17	59 66	38 61	02 10	86 10	51 55	92 52
74 49	04 49	03 04	10 33	53 70	11 54	48 63	94 60	94 49
94 70	49 31	38 67	23 42	29 65	40 88	78 71	37 18	48 64
22 15	78 15	69 84	32 52	32 54	15 12	54 02	01 37	38 37
93 29	12 18	27 30	30 55	91 87	50 57	58 51	49 36	12 53
45 04	77 97	36 14	99 45	52 95	69 85	03 83	51 87	85 56
44 91	99 49	89 39	94 60	48 49	06 77	64 72	59 26	08 51
16 23	91 02	19 96	47 59	89 65	27 84	30 92	63 37	26 24
04 50	65 04	65 65	82 42	70 51	55 04	61 47	88 83	99 34
32 70	17 72	03 61	66 26	24 71	22 77	88 33	17 78	08 92
03 64	59 07	42 95	81 39	06 41	20 81	92 34	51 90	39 08
62 49	00 90	67 86	93 48	31 83	19 07	67 68	49 03	27 47
61 00	95 86	98 36	14 03	48 88	51 07	33 40	06 86	33 76
89 03	90 49	28 74	21 04	09 96	60 45	22 03	52 80	01 79
01 72	33 85	52 40	60 07	06 71	89 27	14 29	55 24	85 79
27 56	49 79	34 34	32 22	60 53	91 17	33 26	44 70	93 14
49 05	74 48	10 55	35 25	24 28	20 22	35 66	66 34	26 35
49 74	37 25	97 26	33 94	42 23	01 28	59 58	92 69	03 66
20 26	22 43	88 08	19 85	08 12	47 65	65 63	56 07	97 85
48 87	77 96	43 39	76 93	08 79	22 18	54 55	93 75	97 26
08 72	87 46	75 73	00 11	27 07	05 20	30 85	22 21	04 67
95 97	98 62	17 27	31 42	64 71	46 22	32 75	19 32	20 99
37 99	57 31	70 40	46 55	46 12	24 32	36 74	69 20	72 10
05 79	58 37	85 33	75 18	88 71	23 44	54 28	00 48	96 23
55 85	63 42	00 79	91 22	29 01	41 39	51 40	36 65	26 11
67 28	96 25	68 36	24 72	03 85	49 24	05 69	64 86	08 19
85 86	94 78	32 59	51 82	86 43	73 84	45 60	89 57	06 87
40 10	60 09	05 88	78 44	63 13	58 25	37 11	18 47	75 62
94 55	89 48	90 80	77 80	26 89	87 44	23 74	66 20	20 19
11 63	77 77	23 20	33 62	62 19	29 03	94 15	56 37	14 09
64 00	26 04	54 55	38 57	94 62	68 40	26 04	24 25	03 61
50 94	13 23	78 41	60 58	10 60	88 46	30 21	45 98	70 96
66 98	37 96	44 13	45 05	34 59	75 85	48 97	27 19	17 85
66 91	42 83	60 77	90 91	60 90	79 62	57 66	72 28	08 70
33 58	12 18	02 07	19 40	21 29	39 45	90 42	58 84	85 43
52 49	40 16	72 40	73 05	50 90	02 04	98 24	05 30	27 25
74 98	93 99	78 30	79 47	96 92	45 58	40 37	89 76	84 41
50 26	54 30	01 88	69 57	54 45	69 88	23 21	05 69	93 44
49 46	61 89	33 79	96 84	28 34	19 35	28 73	39 59	56 34
19 65	13 44	78 39	73 88	62 03	36 00	25 96	86 76	67 90
64 17	47 67	87 59	81 40	72 61	14 00	28 28	55 86	23 38
18 43	97 37	68 97	56 56	57 95	01 88	11 89	48 07	42 60
65 58	60 87	51 09	96 61	15 53	66 81	66 88	44 75	37 01
79 90	31 00	91 14	85 65	31 75	43 15	45 93	64 78	34 53
07 23	00 15	59 05	16 09	94 42	20 40	63 76	65 67	34 11
90 08	14 24	01 51	95 46	30 32	33 19	00 14	19 28	40 51
53 82	62 02	21 82	34 13	41 03	12 85	65 30	00 97	56 30
98 17	26 15	04 50	76 25	20 33	54 84	39 31	23 33	59 64
08 91	12 44	82 40	30 62	45 50	64 54	65 17	89 25	59 44
37 21	46 77	84 87	67 39	85 54	97 37	33 41	11 74	90 50

This table is extracted from the first part of Table 8: Random Sampling Numbers, in D. V. Lindley and J. C. P. Miller, *Cambridge Elementary Statistical Tables*, Cambridge 1953.

100. If a sample of ten items were to be picked then numbers 20, 74, 94, etc. to 04, 32 on that list would form the sample. Again, if the full list were made up of almost 10,000 items then the first *four* columns would be used, again the 0000 representing 10,000. Perhaps in this case a sample of 100 items would be decided upon. The first of these would then be number 2,017 on the full list, the next would be number 7,449, the next number 9,470 until 100 items had been chosen. In this way no bias would be introduced into the choice, for—to quote from the source for the numbers in Table XIV—'Each digit is an independent sample from a population in which the digits 0 to 9 are equally likely, that is, each has a probability of 1/10.' Also, provided that the sample is not so small that it cannot incorporate the full range of conditions in the population, a choice such as this should provide a balanced cross-section of the population conditions—unless, of course, there are really extreme conditions which occur very infrequently. If this is known or found to occur then a rather different method of choosing a sample must be used, as will be outlined below.

In the examples just considered the total population came to the same number as the possibilities involved in the number of digits. It is more often the case that this is not so. For example, the population may total 2,000 items, and therefore four digits must be used for the random numbers. When this happens there are two possible ways of using the random numbers. One method is simply to accept the random numbers which are obtained up to 2,000 and reject (i.e. ignore) those numbers which are obtained between 2,001 and 9,999, carrying on with this until the sample of 100 items is obtained between 1 and 2,000. This can be quite a lengthy process, a very high rejection rate being likely in this example. Instead it is possible to rephrase the numbers above 2,000 as repeats of the 1 to 2,000 series, i.e. numbers 2,001 to 4,000; 4,001 to 6,000 etc. can each be taken as a fresh series of values of 1 to 2,000. Thus all the numbers are used and much time is saved. Another occasion when the renumbering of data is convenient is if the data are available in a series of groups yet it is desired to obtain an overall sample rather than a sample of each group. For example, data may be available concerning the numbers of inhabitants in a large number of settlements. One group of these settlements may be small villages and are returned as such. Another group also returned separately may consist of large villages, another of small towns, and yet another of larger towns. Although it is possible to consider such data in a different way, as will be outlined below, it is also possible to take one sample at random from the whole of the settlements together. To use the table of random numbers in such a case, the numbers must be made to run *consecutively* through the whole population. So if the first group consists of 255 settlements these can be numbered 1 to 255; if the second group contains 176 items these should be renumbered 256 to 431; if the third group consists of 87 values then these should be renumbered 432 to 518; while the fourth group, totalling 18 values, would become 519 to 536. A random sample can then be obtained in the way outlined above.

Apart from this selection from data set out in list form, random sampling methods and techniques can also be applied to data which have an *areal* distribution. In many geographical problems the drawing of samples from within data distributed in space is an essential part of the analysis of the characteristics and qualities of those data. It is often in studies of this sort that there is a great temptation to *select* the items which are to form the sample. For example, in a study of agriculture, 'type' farms are selected for more detailed study because they are known or assumed to represent certain characteristics, or because they are farms for which extra or more accurate information is available. Although this may well give a clear picture of a particular farm, it does not allow generalizations to be made about farming in the area as a whole except by subjective extrapolation. With an experienced and highly qualified research-worker this may be done with a very high degree of accuracy and validity. Any errors that are introduced, however, may be obscured by the treatment, while every worker in the field could very easily arrive at a different answer from every other one as a result of differences inherent in the approach adopted.

Areal sampling by random numbers requires that first of all the area under study should be 'gridded'. In many cases such a grid is already available from the base maps for the area. The grid, whether already on the map or added afterwards, can then be numbered as is the National Grid on the Ordnance Survey maps of Great Britain, i.e. from west to east and from south to north, so that numbers in both directions are at zero in the south-west corner of the area, and increase steadily eastwards and northwards. These numbers can either be made to apply to a grid line or to the space between two grid lines. Which is chosen depends on whether the aim is to sample various *points* or various small *areas*. For example, if it is desired to choose a series of farms for study a 'point sampling' would be necessary. If there were no more than 100 grid lines in each direction then the first four lines of digits in the table of random sampling numbers could be used (or, to yield a finer net, 6-figure groups could be used in the same way as 6-figure grid references). If only four values are used, the first two of these would give the 'easting', i.e. the number of grid lines east of the 'point of origin' in the south-west corner, while the second pair of digits would give the 'northing' from this point. The point thus arrived at will then designate the farm to be included in the sample (Fig. 26*a*). The farm can be specified either as the one in whose land this point lies, or the one whose farmhouse lies nearest to this point. The former will tend to give an over-representation of the number of large farms, but a true representation of the amount of land held by large farms; the latter will tend to over-represent the small farm in terms of the amount of land that falls under small farms, but to give a true representation in terms of the number of small farms. In such a case some 'stratification' of the sample (to be discussed below—p. 113) may be desirable, but the principle remains the same.

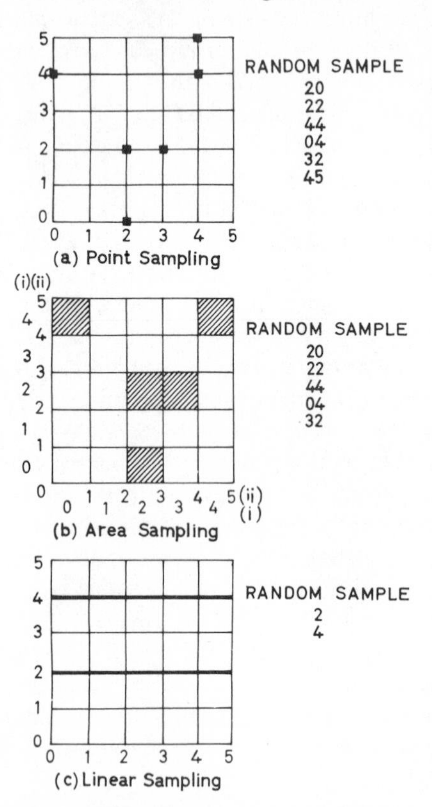

Figure 26. Methods of random sampling for an areal distribution

If instead of farms it is land use for which the sampling is being carried out, then a method choosing a series of small areas might be preferred. In this case the numbering of the grid could apply to the space *between* the grid lines (Fig. 26*b*(i)). Again the table of random sampling numbers would be used and a sample of small areas, within which land use could be plotted rapidly, would be provided. Another means of choosing the areas would be to keep the numbering to the grid lines, and to choose the square to, say, the north-east of the sample point as the sample area (Fig. 26*b*(ii)). Yet a further possibility of areal sampling is by *line samples* (Fig. 26*c*), this proving invaluable for use with binomial distributions such as the irrigation problem considered on pp. 101-103. In this, only eastings *or* northings are needed, and the grid line from this point forms the sample item. Along this line the *distances* possessing or not possessing given characteristics, e.g. irrigation, are measured, these values providing the sample data.

In all these cases it has been implicitly assumed that the overall area is rectangular in shape and that the grid system therefore can fit it exactly. Often this is not so, especially if the overall area is some administrative unit. Even so, a rectangular grid should still be used, ensuring that it provides a full cover for the area. Then if any of the co-ordinates provided by the random numbers lie outside the area under study they should be rejected, as was done with those numbers which fell beyond the limits of listed data (p. 106). Aso, of the three basic methods indicated in Fig. 26, and described in the foregoing pages, the sample based on areas (Fig. 26*b*) clearly includes a larger total area in the sample, but it involves much more work in plotting and calculation. The point sample (Fig. 26*a*) gives a relatively thin sample on the ground, but for the work involved the returns are often high. As for line sampling (Fig. 26*c*), it gives a coverage much closer to that obtained by sampling small areas. The labour is relatively easy from maps, but it can involve innumerable practical problems in the field.

By these various methods, which differ but little from each other, a sample that is strictly random can be obtained from any population, whether this be in the form of a list or of an areal distribution. The purpose of such sampling may simply be to choose certain units for study, these then being described and explained. This, however, is largely a waste of the techniques of sampling, for the data provided by the sample allow further conclusions to be drawn concerning the whole population. The mean and standard deviation of the sample can be obtained in the ways outlined earlier, and from these the sampling standard error can be calculated. Due allowance must here be made for the size of the sample or for the size of the population, again in the ways that have already been explained. In other words, the methods that have previously been considered in Chapter 6 are directly applicable to sample data obtained by the random sampling methods just outlined.

A specific example of this may prove of help. Suppose that it were desired to estimate the relative balance of various types of land-use over the part of Britain represented on the L.U.S. Sheet 13, Kirkby Stephen and Appleby. This could well be done by a spot sample, as is indicated in Fig. 27. In this example a sample of 100 units was taken, each being obtained by a 6-figure grid reference picked from a table of random numbers. Fig. 27 both locates the sample points and shows the land-use (arable, grassland, woodland or moorland) at these points when the land-use survey was made. Also, the overall distribu.ion of moorland is shown, with which the random sample can be compared. From the sample points the following frequencies were obtained which, as the sample was of 100 units, also represent percentages.

Arable	Grassland	Woodland	Moorland	Total
8	31	6	55	100

Furthermore, by using the formula on p. 102, $\left(\sqrt{\frac{p\%.q\%}{n}}\right)$, the standard error for each of these estimates can be calculated. Thus for arable the S.E.

$$= \sqrt{\frac{8 \times 92}{100}} = \sqrt{\frac{736}{100}} = \sqrt{7{\cdot}36} = 2{\cdot}7\%$$

so that the limits of the true percentage of the area under arable (with a 95% probability of being correct) are

$8\% +/- 2$ S.E. $= 8 +/- 2(2{\cdot}7) = 8 +/- 5{\cdot}4 = 2{\cdot}6\%$ to $13{\cdot}4\%$

The standard error for the other three types of land-use are:

	S.E.	Limits of true percentage (at 95% probability)
grassland	$= \sqrt{\frac{31 \times 69}{100}} = \sqrt{21{\cdot}4} = 4{\cdot}63\%$	$31 +/- 9{\cdot}26 = 21{\cdot}74\%$ to $40{\cdot}26\%$
woodland	$= \sqrt{\frac{6 \times 94}{100}} = \sqrt{5{\cdot}64} = 2{\cdot}4\%$	$6 +/- 4{\cdot}8 = 1{\cdot}2\%$ to $10{\cdot}8\%$
moorland	$= \sqrt{\frac{55 \times 45}{100}} = \sqrt{24{\cdot}75} = 4{\cdot}97\%$	$55 +/- 9{\cdot}94 = 45{\cdot}06\%$ to $64{\cdot}49\%$

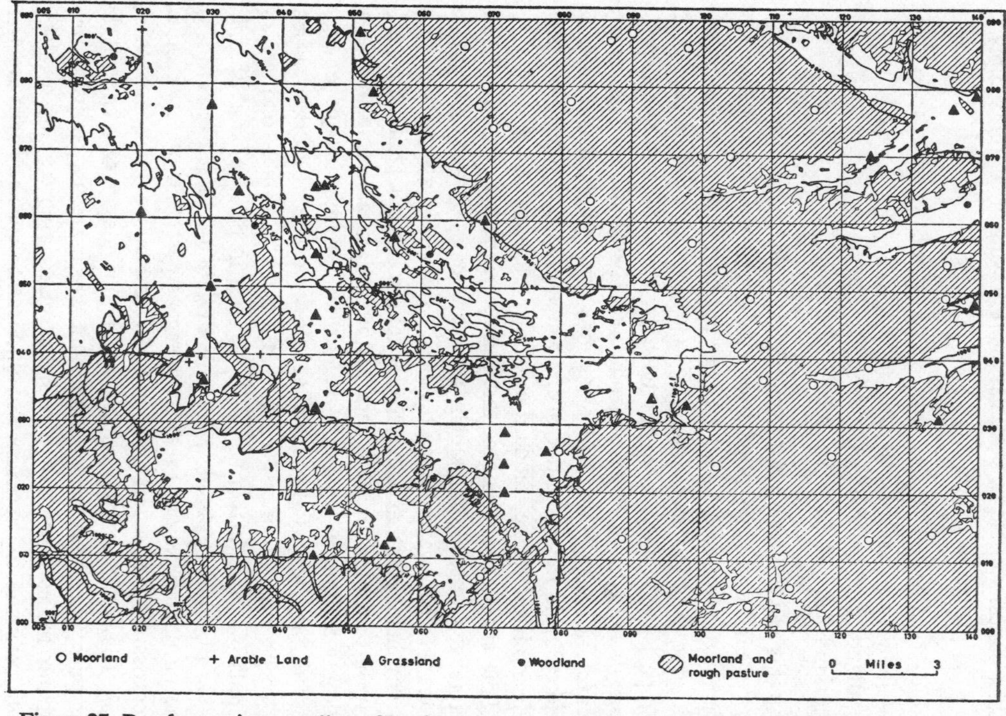

Figure 27. Random point sampling of land-use over L.U.S. Sheet 13, Kirkby Stephen and Appleby

It would, of course, also be possible to translate these percentages into absolute values for the area under review. As the overall area is 532 sq. miles, it could be said that the following areal limits apply at the 95% probability level

arable	14	to	71	sq. mls.	(42·5 sq. mls. from sample)
grassland	116	to	214	sq. mls.	(165 ,, ,, ,, ,,)
woodland	6·5	to	57·5	sq. mls.	(32 ,, ,, ,, ,,)
moorland	240	to	345	sq. mls.	(292·5 ,, ,, ,, ,,)

In this and similar ways it would be possible to assess the land-use, agricultural economy, population, industrial development or any other feature of each of a series of administrative units, the values being required for later comparison (see Chapters 8–10). Each unit, be it parish or county, could be studied by means of a random sample, either from a list of data or from areal distributions. The sample means and sample standard deviations thus obtained can then be used as being representative of the whole unit, once they have been duly modified by the standard error or multiples thereof.

True random sampling of this sort is frequently possible in geographical problems, but equally there are many occasions when this is not so. The most common reason for this is that all too often only *part* of the population data is available. For example, rainfall records may only exist for some 30 years; historical data concerning medieval land-use may be only partially extant; data on industrial production or trade may be partially unavailable for security or business reasons. In such cases the total *available* data is but a *sample* of the total popuiation, and furthermore it is, at least in part, a *biased* sample. So, in the above examples, the rainfall data are biased in favour of one particular period, usually the recent past; the preservation of records may itself reflect some aspect of land tenure which encouraged maintenance of records and this land tenure may in turn control the land-use; the industrial or trade data which are unobtainable may fall into this category just because they are so important, the available data referring to markedly less important aspects. There are thus severe limitations in employing such data as samples from which characteristics of the total population can be assessed. Whether or not this can be done legitimately can often only be decided in the light of other information. For example, if it is known that no significant climatic change has taken place over a prolonged period of time, then the 30 years of records can well be employed as a random sample and assessments based on them accordingly. Again, if it is known that the historical data are all for one homogeneous type of land tenure and that the records are for areas distributed fairly uniformly over the total area, then it may be reasonable to use these records as a sample (almost random in character) of land-use under a given land-tenure system. In this connection, whether or not these records are distributed reasonably in relation to quality of land, for example, could first be tested by the χ^2 Test, which will be presented in Chapter 10. In all these cases, of course, it would be quite legitimate to sample *from* the available data, provided always that any conclusions are only related to these available data and *not* to the total population (unless other evidence, as suggested above, also exists).

Methods of Stratified Sampling

At times, however, it may be more valuable to analyse a body of data in a rather more complex form. In the example on p. 106 to illustrate the consecutive numbering of items for random sampling, the data were stated to fall into several groups. These data concerned settlements that were grouped according to whether they were small villages, large villages, small towns or large towns. In such a case it may be desirable to assess mean conditions, etc. not only for the overall body of data but also for the individual groups separately. Such a grouping of the data, with a sample picked from each group, gives rise to a *stratified sample*, and each group that is sampled is referred to as a *stratum*, i.e. the data, and also the sample, are divided into layers or strata. The analysis of such a stratified sample proceeds in the same way as in the examples outlined earlier, only in this case each *stratum* is sampled by a random sample. To begin, it can be assumed that the proportion of each stratum forming the sample is the same in each case, i.e. that there is a uniform 'sampling fraction'. Random sampling of the total population will yield a close approximation to this, for a random sample tends to select a number of items in each stratum proportional to the size of that stratum. The data can be set out as in Table XV, and it could consist of any one of a variety of aspects of these settlements. Here it can be simply the number of garages serving the settlements, and the sampling fraction (f) may be taken as 1/10.

The number of units in the sample, i.e. the number of settlements studied, is 10% of the total number of units for each stratum. These values are taken to the nearest whole number upwards, to ensure that at least 10% are included. As a result, the *estimates* of the number of units differ from the known number, and the working that follows is related to these estimates—in many studies actual numbers are, in fact, not known. The number of units in the sample is shown in column (*b*). The number of garages serving each settlement is obtained, and the total of such garages for each stratum sample is

Table XV

Tabulation for stratified random sampling with uniform sampling fraction

Strata	No. of units, i.e. settlements, in sample	Sample total of garages	Sample mean of garages per unit	Total no. of units in strata	Estimated total of garages
(*a*)	(*b*)	(*c*)	(*d*) c/b	(*e*) $b.rf$	(*g*) $e.d$
(i) small villages	26	39	1·5	260	390
(ii) large villages	18	36	2·0	180	360
(iii) small towns	9	90	10·0	90	900
(iv) large towns	2	120	60·0	20	1,200
overall values	55 Σb	285 Σc	5·18 $\Sigma c/\Sigma b$	550 Σe	2,850 Σg

entered in column (*c*). The sample mean for each stratum is then obtained by $\frac{\text{column }(c)}{\text{column }(b)}$ and entered in coiumn (*d*). These values allow an estimate to be made of the total number of units in each stratum (although in the present case this is already known), and also an estimate of the total number of garages in each stratum. These values are entered in columns (*e*) and (*g*) respectively. Moreover, estimates can be made concerning the overall body of data. Thus the estimated overall average number of garages per settlement is obtained by $\frac{\Sigma c}{\Sigma b}$, which in this case is $\frac{285}{55} = 5{\cdot}18$. The overall population total, i.e. the total number of garages, can be estimated by multiplying the sample total by the *raising factor* (rf). This latter is the inverse of the sampling fraction (f), such that in this case with $f = 1/10$ then $rf = 10$. This method (i.e. $\Sigma\, c.rf$) would give an estimate of the overall population total of $285 \times 10 = 2{,}850$. If, on the other hand, the actual total number of units is already known, then the overall population total can also be obtained by multiplying this value of the units by the estimated overall mean. The fact that the sampling fraction will not be *exactly* the same in every stratum means that the answer in this case will differ slightly from 2,850, which was obtained by the standard method.

These estimated means and totals, whether they be for strata or the full body, are based on samples and therefore it is necessary to calculate their standard errors. To do this, the standard error must be obtained for each stratum separately (this is usually required as part of the study anyway) and then the standard error for the overall values obtained from the strata values. The calculation of the stratum standard error is carried out in the way outlined for random samples, bearing in mind the need to make the 'best estimate' of the standard deviation (p. 92) and to use Student's t instead of the normal distribution for assessing the limits of the mean when the sample size is small (p. 94). Also, in a study such as this, the population is not one of infinite size even theoretically, but rather it is a *finite population*. For this reason the error involved in sampling will probably be less than in the case of an infinitely large population. Therefore it can be regarded as a 'small' population, and the correction for this—which was indicated on p. 97—can be applied to the calculations of the standard error. This reinforces one of the major advantages of stratified, as distinct from unstratified, sampling—the production of smaller standard errors.

The calculation of the standard error for each stratum therefore proceeds as follows. First the best estimate of the variance of the data is made by the use of the formula

$$\hat{\sigma}^2 = \frac{\Sigma\,(x - \bar{x})^2}{n - 1}$$

This is then adjusted, because of the finite nature of the population, by multiplying it by $(1 - f)$. Then, to obtain the standard error this value is divided by the number of items in the sample and the square root found. Thus the standard error is calculated by the *third* form of the formula set out on p. 97 for use with small (or here, finite) populations, i.e. the standard error for each stratum is obtained by

$$\text{S.E. } \bar{x} = \sqrt{\frac{\hat{\sigma}^2}{n}.(1 - f)}$$

In the example introduced above, this would yield the following values.

Strata	Sample mean	Best estimate of st. dev.	Calculation of standard error of the mean
(i)	1·5	0·5	$\sqrt{\frac{0{\cdot}25}{26} \times 0{\cdot}9} = \sqrt{0{\cdot}00865} = 0{\cdot}09$
(ii)	2·0	0·6	$\sqrt{\frac{0{\cdot}36}{18} \times 0{\cdot}9} = \sqrt{0{\cdot}018} = 0{\cdot}13$
(iii)	10·0	3·0	$\sqrt{\frac{9}{9} \times 0{\cdot}9} = \sqrt{0{\cdot}9} = 0{\cdot}95$
(iv)	60·0	10·0	$\sqrt{\frac{100}{2} \times 0{\cdot}9} = \sqrt{45{\cdot}0} = 6{\cdot}7$

These standard errors must then be applied to the strata sample means so that the limits of the strata true means can be assessed with given probabilities. In this connection it must be remembered that with a small sample the values for the normal distribution must not be used but rather Student's t distribution must be introduced (p. 94). In the present example the third and fourth strata are represented by only small samples, and therefore this adjustment must be made in these cases. The limits of the true means for the several strata, at a 95% level of probability, are therefore as follows:

(i) $\bar{X} = \bar{x} +/- 2.\text{S.E.} = 1{\cdot}5 +/- 2(0{\cdot}09) = 1{\cdot}5 +/- 0{\cdot}18$
$= 1{\cdot}32$ to $1{\cdot}68$

(ii) $\bar{X} = \bar{x} +/- 2.\text{S.E.} = 2{\cdot}0 +/- 2(0{\cdot}13) = 2{\cdot}0 +/- 0{\cdot}26$
$= 1{\cdot}74$ to $2{\cdot}26$

(iii) $\bar{X} = \bar{x} +/- t.\text{S.E.} = 10{\cdot}0 +/- 2{\cdot}3(0{\cdot}95) = 10{\cdot}0 +/- 2{\cdot}19$
$= 7{\cdot}81$ to $12{\cdot}19$

(iv) $\bar{X} = \bar{x} +/- t.\text{S.E.} = 60{\cdot}0 +/- 12{\cdot}71(6{\cdot}7) = 60{\cdot}0 +/- 85{\cdot}16$
$=$ nil to $145{\cdot}16$

The accuracy of the estimates of the true means thus vary markedly between the strata, for as was indicated earlier (p. 89) the accuracy of a sample estimate is controlled not by the proportion of the population that it forms but by the number of items in the sample itself. This problem will be taken up again later.

Further problems are the calculation of the standard errors of the overall sample mean and of the estimate of the overall population total. In both these cases, calculations must first be applied to each of the strata to obtain the 'sample sum of the squares', i.e. that value which, when divided by n, gives the variance. This is obtained as follows. The best estimate of the variance for a finite population is $\hat{\sigma}^2.1-f$, a procedure similar to that employed on pp. 96–97 to calculate the standard error of a sample from a finite population. To obtain the best estimate of the sample sum of the squares this variance must be multiplied by the number of occurrences, i.e. by n (see the method of calculating the variance, p. 24), i.e. it is

$$n.\hat{\sigma}^2.1-f$$

This is the requisite formula for obtaining the best estimate of the sample sum of the squares for each stratum. These separate stratum values must then be summed to give the overall sample sum of the squares

i.e. $\Sigma\,\hat{\sigma}^2.n(1-f)$

From this value the standard deviation and standard error of the mean can be readily calculated. Thus the standard deviation of the overall sample mean involves dividing the sample sum of the squares by the number of occurrences in the sample, and finding the square root (p. 24),

e. $\sqrt{\dfrac{\Sigma\,\hat{\sigma}^2.n(1-f)}{n}}$

If this is then put above $\sqrt{n}$ the standard error of the overall sample mean is obtained (p. 88), so that this standard error is written as

$$\frac{\sqrt{\dfrac{\Sigma\,\hat{\sigma}^2.n(1-f)}{n}}}{\sqrt{n}} = \sqrt{\frac{\dfrac{\Sigma\,\hat{\sigma}^2.n(1-f)}{n}}{n}} = \sqrt{\frac{\Sigma\,\hat{\sigma}^2.n(1-f)}{n^2}}$$

$$= \frac{\sqrt{\Sigma\,\hat{\sigma}^2.n(1-f)}}{n}.$$

It is this latter form, i.e. $\dfrac{\sqrt{\Sigma\,\hat{\sigma}^2.n(1-f)}}{n}$, which represents the most convenient expression for the standard error of the overall sample mean. In practice it is an easy formula to apply, as can be seen in Table XVI in relation to the present example. It will be seen that values for $\hat{\sigma}$ have here been assigned to the samples for each stratum, while the values for n and f are the same as were used in Table XV.

Table XVI

Calculation of the standard error of the overall sample mean for stratified random sampling with uniform sampling fraction

Strata	$\hat{\sigma}$	$\hat{\sigma}^2$	n	f	$n(1-f)$	$\hat{\sigma}^2.n(1-f)$
(i)	0·5	0·25	26	0·1	26 × 0·9 = 23·4	0·25 × 23·4 = 5·85
(ii)	0·6	0·36	18	0·1	18 × 0·9 = 16·2	0·36 × 16·2 = 5·85
(iii)	3·0	9·0	9	0·1	9 × 0·9 = 8·1	9·0 × 8·1 = 72·90
(iv)	10·0	100·0	2	0·1	2 × 0·9 = 1·8	100·0 × 1·8 = 180·00
						$\Sigma \hat{\sigma}^2.n(1-f)$ = 264·60

$$\text{Standard error of the overall sample mean} = \frac{\sqrt{\Sigma\,\hat{\sigma}^2.n(1-f)}}{n}$$

$$= \frac{\sqrt{264{\cdot}60}}{55} = \frac{16{\cdot}24}{55} = 0{\cdot}296$$

As the overall sample mean is 5·18 garages per settlement, the true overall mean (with a probability of 95%) is 5·18 +/− (2 × 0·296) = 5·18 +/− 0·592 = 4·588 to 5·772, i.e. between 4·6 and 5·8 approximately.

To find the standard error of the estimate of the overall population total (i.e. the estimated total number of garages), this standard error of the overall sample mean must be modified. In effect, if this value of 0·296 represents the standard error of the average number of garages for each (i.e. *one*) settlement, then it must be multiplied by the total number of settlements to give the standard error of the total number of garages. This means that it must be multiplied by n to give the standard error for the *sample* total, and then by rf (the raising factor) to convert this to the standard error of the population total. The necessary formula is therefore:

$$rf.n.\frac{\sqrt{\Sigma\,\hat{\sigma}^2.n(1-f)}}{n} = rf.\sqrt{\Sigma\,\hat{\sigma}^2.n(1-f)}$$

The major component of this—$\sqrt{\Sigma\,\hat{\sigma}^2.n(1-f)}$—has already been calculated with the standard error of the overall sample mean above, and in the present example this is 16·24. As the raising factor is 10 (p. 115), the standard error of the overall population total is simply $16{\cdot}24 \times 10 = 162{\cdot}4$. The true population value therefore lies, with a probability of 95%, within the limits of +/– 325 of the estimated value of 2,850 (p. 115), i.e. between 2,525 and 3,175. If, as in the present case, the actual total number of items is known, this standard error can also be calculated by multiplying the standard error of the sample mean directly by the number of items, i.e. $0{\cdot}296 \times 536 = 158{\cdot}656$. This can then be applied to the estimate of the population total made from the true number of items (i.e. an estimate of 2,685), in which case the true population total will lie between 2,685 +/– 317, which is between 2,268 and 3,002, again with a 95% probability. The overlap between these two definitions of the limits of the true overall population total is such that both of the estimates (2,850 and 2,685) are clearly reasonable ones.

Finally, by a similar method the formula for calculating the standard error of the stratum sample mean can be converted to the standard error of the stratum population total. Thus the standard error of the sample mean $\sqrt{\frac{\hat{\sigma}^2}{n}.(1-f)}$ (p. 116) is multiplied by n to give the standard error of the stratum sample total and by the raising factor rf to yield the standard error of the overall stratum total (p. 118),

$$\text{i.e. } n.rf.\sqrt{\frac{\hat{\sigma}^2}{n}.1-f} = n.rf.\frac{\sqrt{\hat{\sigma}^2.1-f}}{\sqrt{n}} = \sqrt{n}.rf.\hat{\sigma}.\sqrt{1-f}$$

$$= rf.\sqrt{\hat{\sigma}^2.n(1-f)}$$

Thus in the case of stratum (i)—small villages (p. 114)—the standard error of the estimated population total of 390 is

$$10\sqrt{0{\cdot}5^2 \times 26 \times 0{\cdot}9} = 10\sqrt{5{\cdot}85} = 10 \times 2{\cdot}42 = 24{\cdot}2$$

The true population total for that stratum, with a 95% probability, therefore lies within the limits 390 +/– 48·4 = 341·6 to 438·4. Similar calculations for the other strata can be made by the reader.

As set out here, an analysis of this sort may appear both complex and confusing. In practice, however, the calculations involved are relatively simple, and reliable values are given for many aspects of the study. The formulae for these, discussed and illustrated in the previous pages, are briefly set out in Table XVII.

Table XVII

Formulae for use with stratified random samples with a uniform sampling fraction

(i) Standard error of the stratum sample mean	$\sqrt{\frac{\hat{\sigma}^2}{n}.(1-f)}$	(pp. 115–116)
(ii) Standard error of the overall sample mean	$\frac{\sqrt{\Sigma\,\hat{\sigma}^2.n(1-f)}}{n}$	(pp. 117–118)
(iii) Standard error of the stratum population total	$rf\sqrt{\hat{\sigma}^2.n(1-f)}$	(p. 119)
(iv) Standard error of the overall population total	$rf\sqrt{\Sigma\,\hat{\sigma}^2.n(1-f)}$	(pp. 118–119)

Thus by studying only 55 out of an actual total of 536 settlements (or an estimated total of 550) it is possible to assess the average number of garages serving small villages, large villages, small towns and large towns respectively; the average number of garages per settlement if differences in size of settlement are ignored; and the total number of garages serving settlements of various sizes and in the whole area under study. All these assessments are set out on p. 114, while these values are all given within specified ranges of probability (pp. 116–119). Similar studies of widely varying characteristics other than garages could also be made from this sample, so that from a relatively small group of settlements a detailed picture could be built up which would apply to the whole range of settlements in the area.

A comparable approach could be applied to the binomial distribution illustrated by the land-use example presented on pp. 110–112. The hundred sample sites can be classified not only in terms of land-use, but also into several strata based on altitude, this yielding the following values:

Ht.	Arable	Grassland	Woodland	Moorland	Total
$<500'$	3	4	1	0	8
$500'$–$1,000'$	5	21	5	10	41
$>1,000'$	0	6	0	45	51
All heights	8	31	6	55	100

The requisite standard errors can then be calculated for any of the land-use categories, both for each stratum (i.e. height range) and for the overall sample. The moorland category can be taken as an example—

Ht.	Frequency of moorland	Sample total n	p	q	S.E. ($\sqrt{npq}$)
<500′	0	8	0	1·0	0
500′–1,000′	10	41	0·244	0·756	2·76
>1,000′	45	51	0·883	0·117	2·29
All heights	55	100	0·55	0·45	4·97

Thus the standard error of the sample frequency is given for each stratum and for the overall sample, and this can be readily converted to a percentage value if it is required. For example, between 500′ and 1,000′ the sample frequency of moorland is 10 out of 41 with a standard error of 2·76. This could equally be expressed as a sample frequency of 24·4%, with a standard error of 6·75% $\left(\text{i.e. } \frac{\text{S.E.}}{n} \times 100\% = \frac{2{\cdot}76}{41} \times 100\%\right)$, so that the true frequency of moorland between 500′ and 1,000′ (at the 95% probability level) lies between 10·9% and 37·9% (see Fig. 27).

Variable Sampling Fractions

It will have been noticed, however, that the degree of accuracy in the estimates varies between the strata. In the study of garages it was rather low for the towns, especially the larger towns, because of the small size of the sample. This can be rectified by ceasing to keep the sampling fraction the same for each stratum. Instead it is possible to use a Variable Sampling Fraction, thus drawing a different proportion from each stratum. It is best to vary the sampling fractions in proportion to the standard deviation of the data in the stratum concerned, but it is not always possible or convenient to calculate this standard deviation accurately and therefore a rough estimate is often made. This may be done simply from the range of values involved, or from the mean values, in each case assuming that as these increase so does the standard deviation. Clearly this does not provide an accurate answer, but it does give the relative order of magnitude in most cases. At times, however, the choice of sampling fraction is based on other criteria. For example, it may be known that in the larger units conditions vary markedly from one to the other, possibly to such an extent that each occurrence is a case in itself. With an extreme situation such as this it may even be necessary to study *every* member of one particular group.

Looking at the example of the numbers of garages per settlement, the means on p. 114 are in a rough proportion of 1 : 1 : 5 : 30, while the standard deviation estimates on p. 116 have a ratio of approximately 1 : 1 : 6 : 20. If the latter is accepted as a closer approximation to the suitable sampling proportions, it may be considered desirable to take approximately 2% samples of strata (i) and (ii), increasing the percentage to about 12% for stratum (iii) and to about 40% for stratum (iv). In practice the actual percentages taken are controlled by the need to obtain a whole number of items in the sample and a convenient size for the total sample. The calculation of the new sampling fractions and raising factors is set out in Table XVIII, where it will be seen that the actual sample size has been rounded upwards from the approximate percentages mentioned above, while in the case of stratum (iii) the size has been increased from 11 to 12 to make a total sample size of 30.

Table XVIII

The calculation of sampling fractions and raising factors for stratified sampling with variable sampling fractions.

Total no. of items (p. 107)	Approx. % sample	No. of items in sample	Sampling fraction $\left(\frac{\text{Sample items}}{\text{Total items}}\right)$	Raising factor $\left(\frac{1}{\text{Sampl. fract.}}\right)$
(i) 255	⩾ 2%	6	0·02353	42·5
(ii) 176	⩾ 2%	4	0·02273	44·0
(iii) 87	⩾12%	12	0·13793	7·25
(iv) 18	⩾40%	8	0·44444	2·25

These values are then entered in Table XIX, together with sample means and standard deviations the same as before, but with the tabulation and calculation related to a variable sampling fraction within a stratified sample. The major difference is that the raising factor must be entered into the table and that strata values must be adjusted by this amount before overall values of the mean and total can be estimated. Moreover these different sampling fractions and raising factors must be introduced into the calculation of the various standard errors.

The retention of sample means and standard deviations the same as with the uniform sampling fraction is deliberate, so that differences in standard errors etc. can be more easily appreciated. In reality, these values would differ at least slightly as normally occurs when a fresh sample is taken.

To obtain the overall average it is necessary to multiply the number of units (column *b*) and the sample total for each stratum (column *c*) by the appropriate raising factor (column *e*). Each of these is then summed (i.e. $\Sigma\, b.e$ and $\Sigma\, c.e$) and then the estimated total ($\Sigma\, c.e$) is divided by the estimated number of units ($\Sigma\, b.e$). This gives the estimated overall population mean, which is here 5·01. Also the estimated overall population total is obtained in this same calculation, being 2,685 in whole numbers.

Table XIX

Tabulation for stratified random sampling with variable sampling fraction

Strata	No. of units in sample	Sample total	Sample mean	Raising factor	Estimated total no. of units	Estimated total	Estimated mean
(*a*)	(*b*)	(*c*)	(*d*)	(*e*)	(*b.e*)	(*c.e*)	(*c.e/b.e*)
(i)	6	9	1·5	42·5	255	383	1·5
(ii)	4	8	2	44·0	176	352	2·0
(iii)	12	120	10	7·25	87	870	10·0
(iv)	8	480	60	2·25	18	1,080	60·0
	30				536	2,685	5·01
	Σb				$\Sigma b.e$	$\Sigma c.e$	

The standard errors for the individual strata are obtained in the same way as was done when the sampling fraction was uniform, although care must be taken to use the requisite sampling fraction in each case. Thus by applying the formula

$$\sqrt{\frac{\hat{\sigma}^2}{n}.(1-f)}$$

and with $f = 0{\cdot}024$, $0{\cdot}023$, $0{\cdot}138$ and $0{\cdot}444$ in strata (i) to (iv) respectively (p. 122), the following standard errors are obtained:

stratum (i) 0·20; stratum (ii) 0·30; stratum (iii) 0·80; stratum (iv) 2·64

On comparing these with those given on p. 116 it will be found that the present values are higher for strata (i) and (ii), but lower for strata (iii) and (iv). As the latter are the ones with the highest mean values, and which on other grounds are probably the more important, such an improvement is valuable.

When calculating the standard error for the overall mean the basic approach is once more the same as in the earlier example, i.e. for each stratum the sum of the squares is obtained by $\hat{\sigma}^2.n(1-f)$. As the sampling fraction varies, however, it is necessary to multiply these in each case by the *square* of the respective raising factor before they are summed. (This factor must here be squared to equate with the 'sum of the squares' which it is modifying, whereas in the case presented on p. 118 it was the standard deviation that was being modified.) This sum is then divided by the estimated overall total number of items (N), instead of the sample number of items (n), before the square root is calculated to yield the standard deviation, i.e.

$$\sqrt{\frac{\Sigma\,\hat{\sigma}^2.n(1-f).(rf)^2}{N}}$$

To obtain the standard error this is then divided by $\sqrt{N}$, and by the same process of cancellation, etc. as on p. 117 the formula for the standard error of the overall sample mean becomes

$$\frac{\sqrt{\Sigma\,\hat{\sigma}^2.n(1-f).(rf)^2}}{N}$$

In the present example the values in Table XX are obtained (for detailed components, see pp. 116, 122 and Table XIX).

Table XX

Calculation of the standard error of the overall sample mean for stratified random sampling with variable sampling fraction

Strata	$\hat{\sigma}^2.n(1-f)$	$(rf)^2$	$\hat{\sigma}^2.n(1-f).(rf)^2$
(i)	1·46	1,806	2,637
(ii)	1·41	1,936	2,730
(iii)	93·10	52·56	4,893
(iv)	444·80	5·06	2,251
		$\Sigma\,\hat{\sigma}^2.n(1-f).(rf)^2 =$	12,511

$$\text{Standard error of the overall sample mean} = \frac{\sqrt{12{,}511}}{536} = \frac{111{\cdot}9}{536} = 0{\cdot}21$$

With these standard errors thus calculated it is possible to establish the limits of the *true* values, and the following list gives them with a probability of 95%. As all the samples are small, however, Student's t distribution has to be used instead of the normal distribution.

Body of data	Sample mean	S.E.	t	$t \times$ S.E.	Limits of true mean (95% probability)
Stratum (i)	1·5	0·20	2·57	0·51	0·99– 2·01
Stratum (ii)	2·0	0·30	3·18	0·95	1·05– 2·95
Stratum (iii)	10·0	0·80	2·20	1·76	8·24–11·76
Stratum (iv)	60·0	2·64	2·36	6·23	53·77–66·23
Total population	5·01	0·21	2·00	0·42	4·59– 5·43

Equally the standard errors of the estimated totals both of the strata and the overall populations can be calculated with little further difficulty. For each stratum the standard error of the estimated total can be obtained by the same formula as on p. 119, i.e. standard error of the stratum population total $= rf.\sqrt{\hat{\sigma}^2.n(1-f)}$. In this case, however, the values for f and rf will be different for each stratum. Thus for stratum (ii) the standard error of the estimated total of 352 would be

$$rf.\sqrt{\hat{\sigma}^2 n.(1-f)} = 44\sqrt{0{\cdot}6^2 \times 4 \times 0{\cdot}977} = 44\sqrt{1{\cdot}41} = 44 \times 1{\cdot}2 = 52{\cdot}8$$

In the case of the overall population total, the standard error is obtained as on p. 118, except that the appropriate raising factor must be applied to each stratum individually, rather than to the sum of the values. It is therefore squared, as in the case on p. 124. So this standard error becomes

$$\sqrt{\Sigma\, \hat{\sigma}^2 . n(1-f).(rf)^2}$$

The components for this are all included in the earlier calculations for the standard error of the overall sample mean (p. 125), so that in the present example it becomes

$$\sqrt{12{,}511} = 111{\cdot}9$$

In this way the limits of the overall population, at the 95% level of probability, are 2,685 +/− 224, i.e. from 2,461 to 2,909.

All these limits, obtained by the formulae shown in Table XXI, are fairly closely defined. The slightly wider limits for the first two strata, as compared to the uniform sampling fraction, are more marked proportionately than in terms of actual values, while the improvement in the degree of reliability of the estimates in strata (iii) and (iv) is most valuable. Furthermore, the overall estimates of both mean and total values are also more closely limited in range, this despite the fact that the total sample consisted of only 30 settlements compared to 55 when using the uniform sampling fraction. This increased degree of accuracy with variable sampling fractions is one of its most valuable attributes and this method of analysis should be used whenever possible.

Table XXI

Formulae for use with stratified random samples with a variable sampling fraction

(i) Standard error of the stratum sample mean	$\sqrt{\frac{\hat{\sigma}^2}{n}.(1-f)}$	(p. 124)
(ii) Standard error of the overall sample mean	$\frac{\sqrt{\Sigma\, \hat{\sigma}^2 . n(1-f).(rf)^2}}{N}$	(p. 124)
(iii) Standard error of the stratum population total	$rf.\sqrt{\hat{\sigma}^2 . n(1-f)}$	(p. 125)
(iv) Standard error of the overall population total	$\sqrt{\Sigma\, \hat{\sigma}^2 . n(1-f).(rf)^2}$	(p. 126)

In fact, it can be extended even further, sub-strata being defined. For example, it would be possible for each of the four strata used above to be sub-divided in terms of areal characteristics, whether these be defined in terms of north–south location, of administrative units, or of any other feature. With increased sub-division, however, it is essential that there be an adequate sample in each sub-stratum, at least if it is desired to calculate the overall error. Innumerable geographical problems, which involve large numbers of items, can be analysed in this way—farms can be grouped into regions (strata) and size (sub-strata) and their characteristics defined by sampling; rivers grouped into length and volume of flow; relief forms grouped in terms of rock lithology and degree of dissection; rainfall data grouped in terms of altitude and location. The possibilities are infinite, and in all cases a relatively close assessment of the characteristics of a large body of data can be obtained by analysing a fairly limited sample, provided that the sampling is organized effectively.

Systematic Sampling

A stratified sample such as this, with random sampling within each stratum and a variable sampling fraction to ensure an adequate coverage of all strata, is probably the most effective way of sampling on this scale. At times, however, it may be desired to adopt a *systematic sampling* technique. By this is meant that items are picked at some regular interval, e.g. every 10th item on a list; every 20th grid square; every 100th line across a map. This is permissible, and provided that there is no periodic repetition of conditions at the same interval as the sample interval, then in general such a sample can be worked as a random sample or as a sample stratified in some predetermined manner. The calculation of sample and population means and totals can be effected as in the case of a random stratified sample. This is perhaps yet more legitimate if it can be argued that the population data, from which the systematic sample is to be picked, are themselves distributed in a random fashion.

At times, a random element is superimposed on a systematic sample. Thus a grid may be located randomly over an area (in terms of its point of origin and its orientation), but the sample points may be at each grid intersection. There is a very wide range of permutations and combinations which may be used, but consideration of them is beyond the scope of such an introductory discussion as this. It is also possible to calculate an approximate standard error for the *strata* values by the same method as was used for a random sample from within a stratum (Table XVII). No fully valid estimate is possible of the standard error of the overall mean or the overall total, however, although various devices allow of a general approximation. For these the reader should turn to one of the more advanced texts on the theme of sampling, as also for any further investigation into sampling possibilities or techniques as a whole, e.g. cluster sampling. These further studies, however, are virtually all based upon the essential foundations of sampling that have been outlined here, and a thorough understanding of these is required before the more advanced methods are consided. Moreover, for a large proportion of the problems that confront geographers the methods already outlined will prove quite adequate.

The particular sampling techniques used, and the detail and complexity of the answers obtained, must always be related ultimately to the problem under study, to the degree of accuracy that is necessary and to the sort of answer that is required. In all such cases, however, the answers in terms of means, standard deviations or totals are only *sample* values. In making estimates of the *true* values from these samples it is necessary to be aware of, and to be able to calculate, the standard error that such sampling introduces, so that the true values can be estimated within given limits. The values obtained in this way may only be required as an indication of the characteristics of that set of data, without further studies being based on such characteristics. More often, however, it is also desired to *compare* the characteristics of different sets of data, so that some judgment can be made concerning their similarities or differences. If the characteristics thus being compared are themselves based upon sample data, from which the true values are estimated, then it is essential that the sampling error be remembered and considered when such comparisons are being made. It is with such problems of comparison, and bearing in mind the various themes which have already been considered in this and earlier chapters, that the following three chapters are concerned.

CHAPTER 8

THE COMPARISON OF SAMPLE VALUES—I

Statistical Significance

So far attention has been mainly concentrated on the primary problem of defining, as briefly and concisely as possible, the conditions presented by a set of data, those data often having been selected by specified statistical methods.

The geographer's interest in quantitative analysis, however, is not limited to such studies of single sets of data, useful and instructive as such analyses may well be. Far more frequently it is necessary to compare one set of values with one or more other sets, as was suggested at the end of Chapter 7. The express purpose of such a comparison is usually either to group together similar sets so as to delimit regions of relatively similar conditions, or to assess the degree of difference between sets of data so that valid comparisons can be made. Very often such comparisons have been but a slight advance on purely subjective assessment, being made merely by a simple inspection of sample mean values—the 'battleship' diagram purporting to show rainfall regime is one of the more common examples of this. Yet it is in this very problem of comparing the degree of similarity or dissimilarity between different sets of data that standard statistical techniques can prove of the greatest assistance to the geographer. With but slight extra labour they can readily provide relatively objective means of analysis which *at least* should prevent conclusions of doubtful validity being drawn and *at best* should enable virtually firm deductions to be made in many cases. Decisions concerning the validity of the difference between various sample values can thus be taken out of the realm of guesswork and brought into that of statistical probability.

The possible methods that can be used for the purpose of comparison are many and varied. Some of these methods can be used only in certain cases, while at other times several of the methods may be pertinent and permissible, and a choice has to be made between them according to ease of computation and the degree of accuracy required. Within the following three chapters several of these various methods will be outlined and each will be used to analyse a number of problems, the examples being deliberately chosen to illustrate the diversity of fields in which the methods may be employed.

The general problem considered in these chapters is one which frequently confronts the geographer, i.e. whether the difference that is apparent between two samples is such that further conclusions can validly be based on this difference or whether the difference is more apparent than real. For example, in a comparative study of two coal-fields (A) and (B) ten pits may be chosen at random (see p. 104) from each field and the production of each pit obtained over a given period. It is necessary to establish whether or not one of these coalfields has a significantly larger production of coal per pit than the other. From this consideration many others would then develop, such as why this difference exists, or its influence on industrial activity or on trade in coal. These later considerations are all dependent on a correct assessment of whether or not the two coalfields differ significantly in terms of production per pit. Is any apparent difference between these two sets of sample data a *statistically significant* difference or is it likely to have been the result of mere chance related to the particular sample of ten pits in each field for which data were obtained?

This phrase—*statistically significant*—will recur frequently in later sections and is explained more fully on pp. 142–146. In general, if a difference is said to be statistically significant this means that it is extremely improbable that such a difference could have occurred by chance. This has two main implications. First it implies that if, instead of the actual values under consideration, other samples were taken of these conditions, or if the full body of data were taken, then it is extremely likely that this difference would *still* be observed—always assuming that the sample being considered is representative of the full body of data, i.e. that it has been chosen by satisfactory methods (see Chapter 7). Second, it implies that if the values recorded in the sets of data being compared were all put together and two samples picked from this grouped collection at random, i.e. by chance, then the difference between these 'chance' samples would be less than the difference between the actual samples being compared. The criteria for deciding on statistical significance in this sense, and the degree of reliability to be placed on such a decision, are varied and will be examined in the succeeding pages.

Parametric and Non-parametric Tests

The type of test or technique that is suitable for effecting a comparative assessment between bodies of data depends largely on the character of the data being studied. Data fall into three main groups —nominal or classificatory; ordinal or ranking; and interval—and the nature of these requires some prior consideration.

(*a*) *Nominal.* This is a group of data which all too often has been assumed by geographers to preclude quantitative testing. Moreover, it is a frequently occurring category of data in geography—the distinction of settlements into Celtic, Anglo-Saxon and Scandinavian origins; the classifying of soils into podzols, brown earths and rendzinas; the distinction of forest, grassland and heath vegetation complexes; the recognition of various tribal, racial or cultural groups; the functional divisions of towns or the land-use divisions of rural areas. None of these carries implications of quantity, not even of relative order of magnitude; they simply refer to categories that are different from one another. Nevertheless, under sampling the various categories may occur with differing degrees of frequency and it is then essential to establish whether such frequency differences are statistically significant or not.

(*b*) *Ordinal.* This is also a very common group of data in geography, in that the *relative* importance (or order of magnitude) of data may be known, even though their absolute values are not. In other words, the data can be ranked, or put in order, either individually or in classes. This may often apply to economic or social data, for example, especially when they are obtained from official bodies such as employment exchanges or the Department of the Environment, which are often precluded by law from making absolute values available. Thus the numbers of people employed by individual firms may vary from one to some high value, but data may only be available in a series of classes (1–20, 21–50, 51–100, etc.). Again, the profitability or costs of certain operations may be defined by firms or farmers as high, medium and low, because they are unwilling to make actual values available. At other times, difficulties of measurement or recording may make the ordinal form of data more convenient, as when classifying river-bed load as coarse, medium and fine, or slopes as steep, moderate and gentle, or soils as acid, neutral and alkaline. In most of these cases, however, there is some implicit underlying continuum in terms of magnitude, the discrete categories being merely a convenient division, and this characteristic is assumed in many of the tests that are available.

(*c*) *Interval.* When not only is the order of magnitude known, but also the actual degree of magnitude as well, then an interval scale exists. This is characteristic of rainfall data, production values, population returns, and many other types of data of geographical relevance. Often, such data are expressed as *ratios*, as when they are converted into percentage values.

The types of statistical tests of significance that can be used vary between these different sorts of data. For nominal data only the simpler tests may be used; for ordinal data rather more sophisticated tests are possible, although the simpler tests for nominal data can also be employed if this is so desired; while for interval data not only are all these techniques available but also a series of others of greater stringency and refinement. A basic distinction can be drawn, however, between those tests which can only be used for interval data, and those which are also suitable for the nominal and ordinal groups too. The former are called *parametric* tests and the latter *non-parametric* tests.

(i) *Parametric tests*

By the word 'parameter' we imply some quality, characteristic or value of the *population data*, not of the sample data. Parametric tests are therefore those that assume certain conditions in the population, and which are relevant only if such assumptions are valid. Most commonly these assumptions are:

(*a*) that the population data are normally distributed;

(*b*) that the observations are independent of each other, i.e. that by picking one of them others are not precluded from consideration, and that the value of any one case does not affect that of others;

(*c*) that the populations being compared have the same variance;

(*d*) that the variables are available on the interval scale.

Thus tests in this category are based on variance, standard deviation, etc. and are tied to the normal frequency distribution. The two most commonly used are the Student's t test, which will be considered later in this chapter, and Snedecor's F test, which will enter into Chapter 9.

(ii) *Non-parametric tests*

These tests do not make a series of assumptions about the population values, save perhaps that of independence, and that there is some underlying continuity in the data. As a result, not only can they be used for data on the ordinal scale, and even on the nominal scale at times, but they are also invaluable for testing interval scale data when either the form of the frequency distribution curve of the population data is not known, or it is known or believed to be skew. In the latter case non-parametric techniques can be used instead of attempting to transform the data to the normal curve in the ways discussed earlier in Chapter 4. Moreover, several of the tests can be effectively employed with only small samples of data, and in this connection it is usually better to use such techniques even with interval scale data. A limited number of examples of these tests will be outlined in Chapter 10.

Dispersion Diagrams

Before considering the strictly parametric tests, it will be of value to outline a graphical technique which partakes of a number of parametric qualities—this is based upon the comparison of dispersion diagrams. A more detailed treatment of the coalfield production problem outlined on p. 130 is very revealing in these terms. A full record for the two samples is set out below, and this gives an opportunity for a more direct assessment of the difference between the two sets of values.

Annual production by sample pits in two coalfields

Coalfield A	Coalfield B
0·25	0·27
0·26	0·28
0·27	0·29
0·27	0·33
0·28	0·34
0·29	0·35
0·32	0·35
0·34	0·38
0·35	0·39
0·37	0·42
Average ($\bar{a}$) = 0·30	Average ($\bar{b}$) = 0·34

This method is best illustrated by plotting the two sets of data (Fig. 29*a*) as dispersion diagrams, and entering the median and quartile values on each. It can be seen that despite the differences in the mean values (the median of A is 0·285 and of B 0·345) there is nevertheless considerable overlap between the two records. Is this overlap so great that there is no significant difference between the records, or is it so slight that it can reasonably be ignored?

Three sets of conditions are regarded as being of diagnostic value and these need to be considered first in general terms before they are applied to this coalfield example. These three sets of conditions are defined simply in terms of the relative positions of the quartile and median values, which can easily be established. In the first case (Fig. 28*a*) the lower quartile of one record (L.Q.$_{1}$) is greater in magnitude than is the upper quartile of the other record (U.Q.$_{2}$), and there is thus a clear space on the diagrams between the ranges of the central 50% of the two sets of data. This relationship can be regarded as indicating a significant difference between the records under analysis. At times, however, this degree of difference is not found, and L.Q.$_{1}$ does not exceed U.Q.$_{2}$. Here a transitional set of conditions can be defined, when the lower quartile of one record is less than the upper quartile of the other but is still greater than the median, while it is the median of the first record which exceeds the upper quartile of the second, i.e. L.Q.$_{1} > M_2$ and $M_1 >$ U.Q.$_{2}$ (Fig. 28*b*). If *both* these conditions are satisfied, then the difference between the two records is *probably* significant but not absolutely so, at least with samples of less than about 40 items. Finally, if either or both of the above conditions do not hold true then no matter what the difference in mean values, it is not safe to assume that the two records are significantly different. Thus flexibility is introduced by this method, a transitional category is defined and a marked degree of difference is required before a fully significant difference is established.

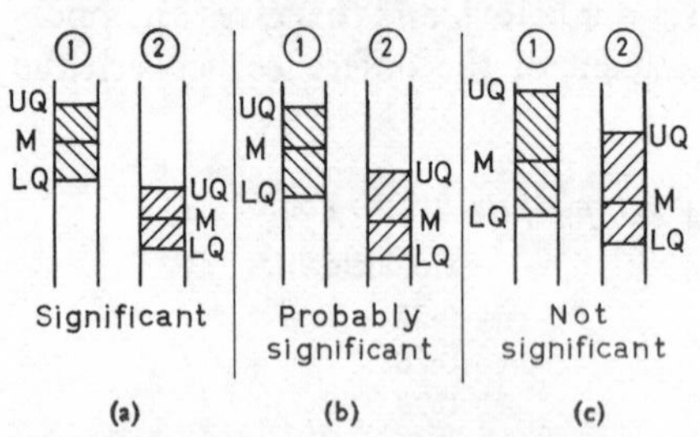

Figure 28. Criteria for the definition of degrees of statistical significance from dispersion diagrams

Armed with this technique, it is possible to return to the coalfield example to assess the degree of significance of the difference between the two sets of data. Figure 29*a* presents the data suitably rearranged. The median of coalfield B is greater than the upper quartile of coalfield A (0·345 as compared to 0·340) and the lower quartile of B is greater than the median of A (0·290 compared to 0·285). Thus the difference between the two coalfields is probably significant but not clearly so, especially as this degree of significance is only just achieved in terms of both criteria considered and the samples are only of 10 items each. Although it is legitimate to continue working on the assumption of a difference in production per pit between the two, one would really like more evidence, i.e. a larger sample, and conclusions should certainly not be pressed too far until such extra evidence has been obtained and analysed.

This simple graphical assessment can be made in any branch of geographical study, and the following two examples provide further illustration of its value. An investigation of low-level cliff remnants around a broad east–west estuary is in progress. On either side of this

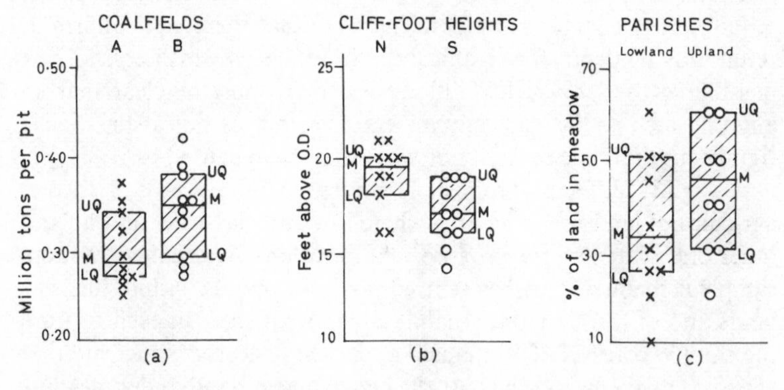

Figure 29. Specific examples of dispersion diagrams used for tests of significance

estuary ten such remnants are found, rising from a wave-cut platform. These are not strictly a random sample, and may in fact be biased in terms of sites favouring preservation. They may be the only data available, however, and as indicated in p. 112, they may therefore be analysed as if they were a random sample, though the results of such an analysis must be used with care. The altitudes of the bases of these cliffs are accurately measured, and the mean value of the cliff base is found for each side of the estuary. On comparison it is found that these mean values differ by 2 ft. if the average is used (17 ft. O.D. on the southern side and 19 ft. O.D. on the northern), and by 2·5 ft. if the median is used (17 ft. O.D. and 19·5 ft. O.D. on the southern and northern sides respectively). Is this small difference in sample mean values merely the result of the limited number of observations, or is there a really valid difference between the two sides of the estuary which merits some explanation? A quick guide to a decision can clearly be made by means of dispersion diagrams, plotted from the following observed values:

heights on S. side (in ft. O.D.)—15, 19, 18, 17, 17, 19, 14, 16, 19, 16.
heights on N. side (in ft. O.D.)—16, 18, 21, 20, 19, 20, 19, 21, 20, 16.

A visual comparison can then be made (Fig. 29*b*), and it is found that lower quartile (N) is greater than median (S), and median (N) is greater than upper quartile (S). In other words, the slight difference in mean height is probably significant, though not conclusively so. The analysis suggests two things. First, more observational data are required (i.e. a larger sample) in the hope that these will confirm or refute this tendency for significance. Second, it is worth considering possible causes for such a difference, for it must be clear that an analysis such as this can only indicate whether or not a statistically significant difference exists, not what may have caused it.

A slightly different problem is presented by a study of former agricultural land-use in an area where lowland clay vales and upland limestone plateaux are in close juxtaposition. A stratified random sample is made of parishes centred upon, or mainly within, the lowlands and of those that are mainly upland parishes. In each stratum the sample consists of 10 items (i.e. parishes). Records indicate that for some given date in the past the percentage of land under meadow in each case was as follows:

lowland parishes (% in meadow)—10, 20, 25, 25, 30, 35, 45, 50, 50, 60.
upland parishes (% in meadow)—25, 30, 30, 40, 40, 50, 50, 60, 60, 65.

A simple calculation shows that the averages for lowlands and uplands differ as between 35% and 45% meadowland, while in terms of median the difference is between 32·5% and 45%. Again the problem is similar; is this difference in sample mean values between two contrasting groups of parishes sufficient to justify an emphasis on contrasting proportions of meadowland as between lowland and upland (or as between clayland and limestone), or is the range of values within each group such that a generalization of that sort is unsound and unjustified? The dispersion diagrams in Fig. 29*c* illustrate the considerable range of overlap between the two sets of data; neither of the criteria for even a 'probably significant' verdict is present. So these data would not justify a claim of a significant contrast between lowland and upland conditions. If a larger sample were studied data justifying such a contrast might well be obtained, but meantime any conclusions claiming a causal relationship between the location of these parish groups and proportions of meadowland would be unsound.

This graphical method of assessing significance is thus simple to apply to a wide range of problems, but it has several limitations. It has only three sets of distinctive conditions without any gradation between them. Moreover, it makes no real allowance for the number of items entering into the computation, which is especially important in the case of such examples as these where the samples are small ones, and the degree of stringency involved in the test for significance could well be intensified. Finally, of course, it is based on the median and quartiles as measurements of mean and deviation values, while it is the arithmetic average and the standard deviation which, as indicated in Chapters 2 and 3, possess the greatest merit in these fields. It is not, in fact, a truly parametric test. It has been used in a number of studies in the past, however, and is still a useful geographical tool.

Standard Error of the Difference

There are two parametric methods which, in large measure, eliminate these disadvantages and although they each involve more computation than do the graphical methods they are, on balance, to be preferred. These two methods will be applied first to the coalfield example, and then to the other problems briefly considered above, and the difference between the methods will thus become apparent. In Chapter 6 the relationship between the mean value of a sample and the true mean value was considered, this relationship (known as the *standard error of the mean*) being expressed by $\frac{\hat{\sigma}}{\sqrt{n}}$ or $\sqrt{\frac{\hat{\sigma}^2}{n}}$ i.e. the best estimate of the standard deviation divided by the square root of the number of items in the sample. The examples considered in the present chapter are also based on samples: the problem is whether or not the differences between these *sample* means are sufficiently great to justify a conclusion that the *true* means also differ significantly.

Thus a comparison is being made between two sample means, each of which has a standard error (S.E.). From the data for the coalfield example set out on p. 133 it can be calculated that coalfield A has a sample mean $\bar{a}$ of 0·30 million tons and a standard error S.E.$_a$ of 0·013 million tons, while for coalfield B the sample mean $\bar{b}$ is 0·34 million tons and standard error S.E.$_b$ is 0·016 million tons.

Coalfield A	Coalfield B
$\bar{a} = 0{\cdot}30$	$\bar{b} = 0{\cdot}34$
$\hat{\sigma}_a = 0{\cdot}042$	$\hat{\sigma}_b = 0{\cdot}05$
$\text{S.E.}_a = \frac{\hat{\sigma}_a}{\sqrt{n_a}} = \frac{0{\cdot}042}{\sqrt{10}}$	$\text{S.E.}_b = \frac{\hat{\sigma}_b}{\sqrt{n_b}} = \frac{0{\cdot}05}{\sqrt{10}}$
$= 0{\cdot}013$	$= 0{\cdot}016$

Both the methods now to be considered utilize these facts, in that both are concerned with assessing, from these data, the standard error of the *difference* between these two sample means, i.e. the standard error of $|\bar{a} - \bar{b}|$. This standard error partakes of the probability characteristics of the normal frequency curve, as did the standard error of the mean (p. 88), so that the probability that the *actual* difference will be more than twice this standard error is about 5%, and that it will be more than three times this standard error is about 0·3%. In other words, if the actual difference between $\bar{a}$ and $\bar{b}$ (in this case 0·04 million tons) is greater than twice the standard error of the difference, then it is unlikely (though not completely certain) that a difference of this size between the two sample means occurred by chance, i.e. the difference is 'probably significant' and it is likely that it would also occur between the true means. If it is greater than three times the standard error of the difference, however, then the difference is almost certainly significant (99·7% certain).

An assessment of the *standard error of the difference* between sample means can thus provide a valuable test of significance. It is based on average and standard deviation values, it allows for the number of items in each sample, and it imposes sufficiently stringent conditions, i.e. odds of at least 19 : 1 before 'probably significant' can be applied, for findings to be accepted with considerable confidence. But how can this standard error of the difference be calculated? The method depends on the fact that the standard error of the mean is a function of the standard deviation, which is itself the square root of the variance (p. 24). So, if the standard error of the sample mean is $\frac{\hat{\sigma}}{\sqrt{n}}$, then the variance of the sample mean is the standard error squared, i.e. $\left(\frac{\hat{\sigma}}{\sqrt{n}}\right)^2$, or more simply $\frac{\hat{\sigma}^2}{n}$. Furthermore, it can be accepted that the variance of the sum of, or the difference between, two sample means is the sum of the separate variances of the two sample means. To put it another way, in adding together two sample means, or in subtracting one from another, each of the values in the calculation has its own standard error and therefore the answer is itself subject to error from *both* sources, i.e. the sample answer, be it sum or difference, is liable to differ more from the true answer than do either of the sample means from their respective true means.

Taking now the variance of the *difference* between two sample means, and applying the rule above:
the variance of the difference between $\bar{a}$ and $\bar{b}$
= the variance of $\bar{a}$ plus the variance of $\bar{b}$

i.e. var. $|\bar{a}-\bar{b}| = \frac{\hat{\sigma}_a^2}{n_a} + \frac{\hat{\sigma}_b^2}{n_b}$ or $\left(\frac{\hat{\sigma}_a}{\sqrt{n_a}}\right)^2 + \left(\frac{\hat{\sigma}_b}{\sqrt{n_b}}\right)^2$

Moreover, as was indicated above, the variance of $|\bar{a}-\bar{b}|$ is the standard error of $|\bar{a}-\bar{b}|$ squared; conversely, the standard error of $|\bar{a}-\bar{b}|$ is the square root of the variance of $|\bar{a}-\bar{b}|$.
i.e. S.E. $|\bar{a}-\bar{b}| =$

$$\sqrt{\frac{\hat{\sigma}_a^2}{n_a} + \frac{\hat{\sigma}_b^2}{n_b}} \quad \text{or} \quad \sqrt{\left(\frac{\hat{\sigma}_a}{\sqrt{n_a}}\right)^2 + \left(\frac{\hat{\sigma}_b}{\sqrt{n_b}}\right)^2}$$

This formula, in either of its forms, can then be applied to the coalfield data, calculating as follows:

S.E. (0·04 mill. tons diff.) =

$$\sqrt{\frac{0\cdot042^2}{10}+\frac{0\cdot05^2}{10}} \quad \text{or} \quad \sqrt{\left(\frac{0\cdot042}{\sqrt{10}}\right)^2+\left(\frac{0\cdot05}{\sqrt{10}}\right)^2}$$

$$=\sqrt{\frac{0\cdot00166}{10}+\frac{0\cdot0025}{10}} \qquad =\sqrt{0\cdot0133^2+0\cdot0158^2}$$

$$=\sqrt{\frac{0\cdot00416}{10}} \qquad =\sqrt{0\cdot000176+0\cdot00025}$$

$$=\sqrt{0\cdot0004}=0\cdot02 \qquad =\sqrt{0\cdot0004}=0\cdot02$$

Returning to the earlier argument that there is only a 5% probability that the actual difference will be as great as twice the standard error of the difference, these two values can now be compared:

actual difference = 0·04 mill. tons
2 S.E. of difference = 0·04 mill. tons

This means that if all the twenty values for the two coalfields were taken together and grouped into two sets of ten purely at random, a difference as great as the one observed, i.e. 0·04 million tons, would occur on no more than 5% of the occasions it was done. So it would occur *by chance* one time in 20, and this 5% (or 2 S.E.) level of probability is taken as the highest chance probability value which can be allowed if the difference is to be described as probably significant. If the difference exceeds two and a half or three times the standard error then it is truly significant, but if it is less than twice the standard error then the difference is *possibly* not significant. It must be stressed that a value of this latter magnitude does not prove that the difference is *not* significant, but rather it indicates that a significant difference has not been adequately proved, and that judgment must be deferred (see pp. 143–144 for further discussion of this in terms of the null hypothesis). What must be borne in mind is that if a difference of this order, i.e. less than 2 standard errors, is obtained, any further deductions based on an assumed difference between the two sets of data may be unsound and are at best unproven.

In the coalfield example this test by the standard error of the difference gives the same answer as did the dispersion diagrams, i.e. probably significant. On the other hand, this present method gives a numerical expression of the degree of significance, and from this it can be appreciated that the difference only just reaches the critical value. On the other hand, obvious critical limits for varying degrees of probability are few, being mainly whole-number multiples of the standard error. Intervening values can be computed but this is a needlessly laborious process.

Student's *t* Test

The best and simplest way to eliminate this difficulty is to apply a more refined technique, though still embodying the standard error of the difference. This technique is known as *Student's t Test* and employs the Student's *t* distribution introduced in Chapter 6 (p. 94). It provides an index—*t*—to represent the relationship between the difference between the means and the standard error of this difference. This index can then be referred to prepared tables or a graph, from which the degree of significance of the difference can be assessed. Student's *t*, the index concerned, is readily calculated as follows:

$$t = \frac{\text{difference between the means}}{\text{standard error of the difference}}$$

$$\text{i.e.} = \frac{|\bar{a} - \bar{b}|}{\sqrt{\frac{\hat{\sigma}_a^2}{n_a} + \frac{\hat{\sigma}_b^2}{n_b}}}$$

So, instead of simply looking to see whether the observed difference is greater than two standard errors, a calculation is made to express exactly how many times greater than the standard error the observed difference really is. In the coalfield example, therefore, it is seen that

$$t = \frac{0{\cdot}34 - 0{\cdot}30}{\sqrt{\frac{0{\cdot}00166}{10} + \frac{0{\cdot}0025}{10}}} = \frac{0{\cdot}04}{\sqrt{0{\cdot}0004}} = \frac{0{\cdot}04}{0{\cdot}02} = 2{\cdot}0$$

It is this value of $t = 2{\cdot}0$ which is checked on tables or graph to find the percentage probability of it occurring by chance. It is here, however, that a more fundamental difference enters into this method as compared with the previous one. On the graph of Student's t (Fig. 30) the co-ordinates are (i) the value of t and (ii) a value representing the number of occurrences on which the comparison is based. This is because in general terms the significance of a given value of t is less the smaller the number of occurrences involved. That is to say, the smaller the samples, the greater the difference between the means must be (i.e. a higher value for t) in order to reach a given level of significance. Once the number of occurrences reaches 35–40 little further change occurs in the required value for t as the number of occurrences increases.

One further point needs stressing, however. As indicated on p. 94 Student's t graph (or table) does not use the exact number of occurrences but instead the number known as the 'degrees of freedom'. By this is meant that number of values that can be assigned arbitrarily, assuming that the sample total and mean remain as they are. So, in the case of coalfield A once nine of the values have been established then the tenth value must follow automatically to yield a sample mean of 0·30 million tons. The same is also true for coalfield B, so for each set of data the degrees of freedom are one less than the number of occurrences, i.e. $(n - 1)$. For the full comparison here, therefore, both values of degrees of freedom must be incorporated, and this can be written as:

degrees of freedom (d.f.) $= (n_a - 1) + (n_b - 1)$
$= n_a + n_b - 2$

In this example degrees of freedom are therefore $10 + 10 - 2 = 18$, this indicating that the value of $t = 2{\cdot}0$ is based on 18 freely varying occurrences (the remaining two occurrences being controlled by the requirements of mean values and the other 18 occurrences).

At this stage it is desirable to introduce the basic reasoning employed in virtually all statistical testing to establish whether or not the apparent difference between samples can be accepted as real.

One begins by setting up an hypothesis that there is, in fact, *no* difference between the bodies of data being compared, even though the sample values might suggest such a difference. Such an apparent difference, it is assumed, results merely from the fact that they are samples, and if other samples had been taken then no such difference, or even a difference in the opposite direction, might well result. The hypothesis, which assumes no real difference between the data, is referred to as the *null hypothesis*, and is usually written as H_o. Statistical testing then consists of deciding:

(*a*) whether one should accept or reject the null hypothesis; if one rejects it, then necessarily the inverse of the null hypothesis (written as H_1) must be accepted, i.e. that there *is* a real difference between the sets of data, as suggested by the samples;

(*b*) the probability that, when rejecting or accepting the null hypothesis, one has unwittingly made an error.

To assist in making such decisions in statistical testing, an index appropriate to the particular test is calculated. Having obtained a diagnostic index (here t) and a measure of sample size (here, degrees of freedom), these are referred to a graph (or table) which contains one or more critical limits. In Fig. 30, for example, four such critical limits are shown. If the lowest of these on the graph, marked as 10%, is taken, then the area below it is classified as the *acceptance zone* and the area above it as the *rejection zone*. These terms indicate whether one accepts or rejects the null hypothesis if the point specified by the t index and the degrees of freedom falls in the relevant zone. As for the percentage value of the critical limit, this shows the *percentage probability of being wrong when one rejects the null hypothesis*. This is referred to as *type I (or α) error*, and it should be noted that in this and all other tests the critical limits given are specified in terms of this type I error. Thus from the example above, degrees of freedom were 18 and $t = 2{\cdot}0$. On entering these on Fig. 30, the point will be seen to fall above the 10% line, i.e. in the zone for rejecting the null hypothesis; moreover, in so rejecting it and accepting the inverse hypothesis that there *is* a difference, there would be no more than a 10% chance of making a mistake.

This chance, however, is usually considered to be too large, and the largest percentage probability of wrongly rejecting the null hypothesis that is usually acceptable is 5%. If this were now taken on Fig. 30 as our critical limit, the conditions given above would fall in the acceptance zone, i.e. one would have to accept that no difference had been established at an acceptable level of reliability (compare the result on p. 135). However, it is no doubt clear to the reader that this verdict could be wrong, i.e. that there exists a *probability of being wrong when accepting the null hypothesis*. This is referred to as *type II* (or β)*error*, and unfortunately there is no simple method of computing this error without considerable labour. It means in practice that if a value falls in the acceptance zone but near to the critical line this should not be interpreted as proving that no difference exists, but rather that one has failed to prove that there is a difference. The Scottish verdict of 'not proven' is the best anology.

The practical means of allowing for (rather than calculating) type II error are twofold.

(*a*) as the value for β increases as that for α decreases, then type II errors will be fewer if the less stringent 5% level for α is taken,

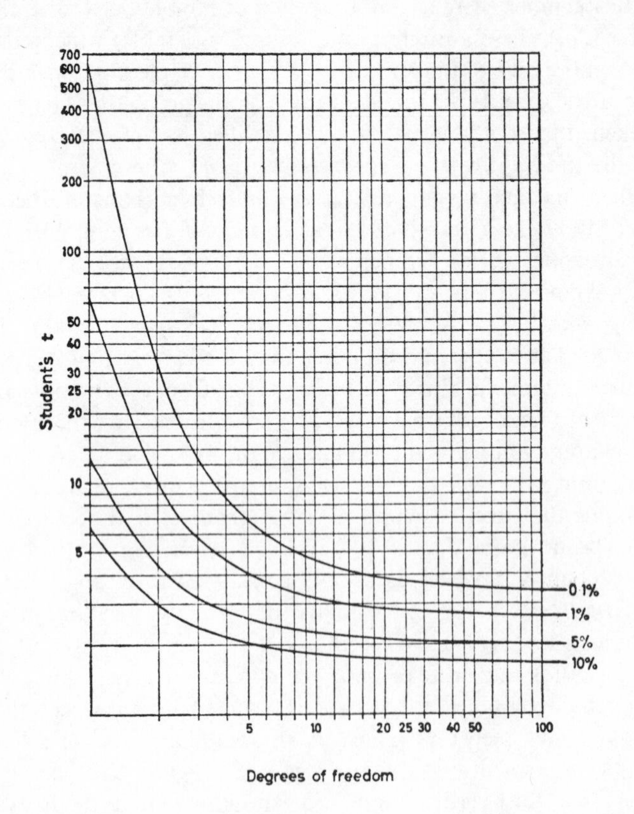

Figure 30. Student's *t* Graph (based on data in D. V. Lindley and J. C. P. Miller *Cambridge Elementary Statistical Tables*, Table 3)

rather than the more stringent 1% or 0·1% levels (these are often written as $\alpha = 0{\cdot}01$ and 0·001 in proportionate terms).

(*b*) Other things being equal, β decreases as the size of sample increases.

Whether or not to bother with these considerations depends on the aim of the analysis. If one is looking for only those differences that are really reliable, then a stringent test for rejection of the null hypothesis is needed, i.e. $\alpha = 0{\cdot}01$ or $\alpha = 0{\cdot}001$. If, on the other hand, one hopes to establish as many differences as possible, and does not wish to accept the null hypothesis wrongly too often, then $\alpha = 0{\cdot}05$ can be used, coupled with as large a sample as possible. What cannot be done, however, is to establish categorically that two sets of conditions are the same because of accepting H_o; all that can be done is to show that differences cannot be established—similarity is then a personal inference, not a statistical result.

The calculations involved in these methods usually follow on the prior computation of average and standard deviation values for other purposes. Furthermore, there is nothing abstruse or difficult in them, for squares and square roots represent the most advanced techniques required. However, the reader may wish to see these methods employed in one or two examples, and it is therefore proposed to rework the geomorphological and agricultural examples previously analysed by graphical methods.

For the comparison of cliff-foot heights on either side of an estuary the data presented on p. 136 can be summarized as follows:

southern side—average ($\bar{x}_1$) = 17; best estimate of S.D. ($\hat{\sigma}_1$) = 1·76; no. of items (n_1) = 10

northern side—average ($\bar{x}_2$) = 19; best estimate of S.D. ($\hat{\sigma}_2$) = 1·826; no. of items (n_2) = 10

The problem is to assess the significance of this difference of 2 ft. between the sample means. As with the coalfields, the comparison of dispersion diagrams in this problem gave a 'probably significant' answer, but as the coalfield example showed, this need not necessarily be the answer given by these other methods. To calculate the standard error of the difference, so as to use it as a test, the following formula is again used:

$$\text{S.E.}\,|\bar{x}_1 - \bar{x}_2| = \sqrt{\frac{\hat{\sigma}_1^2}{n_1} + \frac{\hat{\sigma}_2^2}{n_2}}$$

$$= \sqrt{\frac{1{\cdot}76^2}{10} + \frac{1{\cdot}826^2}{10}} = \sqrt{\frac{3{\cdot}10}{10} + \frac{3{\cdot}33}{10}}$$

$$= \sqrt{0{\cdot}310 + 0{\cdot}333} = \sqrt{0{\cdot}643} = 0{\cdot}802$$

Twice the standard error thus gives a value of 1·604 and three times the standard error one of 2·406. The actual difference being 2·0, i.e. between two and three standard errors, it is clearly a probably significant one.

If this standard error of the difference is employed in Student's t Test the same general answer is obtained. A null hypothesis is postulated, i.e. that there is no difference in height between the two sides of the estuary. Then:

$$t = \frac{|\bar{x}_1 - \bar{x}_2|}{\sqrt{\frac{\hat{\sigma}_1^2}{n_1} + \frac{\hat{\sigma}_2^2}{n_2}}} = \frac{2{\cdot}0}{0{\cdot}802} = 2{\cdot}49$$

and degrees of freedom $(n_1 + n_2 - 2)$ are 18.

If the critical level is taken as the 5% line (i.e. $\alpha = 0{\cdot}05$) then these values fall in the rejection zone for the null hypothesis. It is therefore possible to accept the inverse hypothesis that there *is* a difference with no more than a 5% chance of being in error. On the other hand, if $\alpha = 0{\cdot}01$ were the critical limit, it would be necessary to accept the null hypothesis, i.e. one had failed to establish a difference. If more detailed tables are used instead of this graph it can be shown that H_0 can be rejected with only a 3% chance of being in error. This is often expressed in such terms as 'the observed difference is significant at the 3% level'.

Another fairly common convention is to describe a difference established with only 1–5% probability of being wrong as *probably significant*, one with 0·1–1·0% error probability as *significant*, and one with an error probability of less than 0·1% as *highly significant*. So in this case the application of these mathematically sounder and more stringent tests not only confirms that the difference observed is probably significant, but also indicates that the difference is much nearer being truly significant than could possibly be assessed by means of the dispersion diagram.

As for the comparison of proportions of meadowland between lowland and upland parishes, even the dispersion diagrams suggested a lack of significance. Some indication of the degree to which the difference fails to be significant could well be of value, however, and it is therefore worth while analysing by these further methods. The data may be summarized as follows:

lowland parishes $\bar{x}_1 = 35$; $\hat{\sigma}_1 = 15{\cdot}8$; $n_1 = 10$
upland parishes $\bar{x}_2 = 45$; $\hat{\sigma}_2 = 14{\cdot}15$; $n_2 = 10$

From these data the standard error of the difference may be readily calculated:

$$\text{S.E.}\ |\ \bar{x}_1 - \bar{x}_2\ | = \sqrt{\frac{\hat{\sigma}_1{}^2}{n_1} + \frac{\hat{\sigma}_2{}^2}{n_2}} = \sqrt{\frac{15{\cdot}8^2}{10} + \frac{14{\cdot}15^2}{10}}$$

$$= \sqrt{\frac{250}{10} + \frac{200}{10}} = \sqrt{25{\cdot}0 + 20{\cdot}0} = \sqrt{45{\cdot}0}$$
$$= 6{\cdot}7$$

As twice the standard error is thus 13·4, and the actual difference is only 10, it is clear that this method too indicates a lack of significant difference. In such cases there is no real need for the t test, for with $t = 1{\cdot}49$ and 18 degrees of freedom, the null hypothesis has to be accepted even if $\alpha = 0{\cdot}1$ (i.e. if the 10% line is taken as critical).

By now the relative advantages and disadvantages of these differing methods must be fairly obvious. For rapid comparison, especially if the data involve numbers which prove awkward for easy calculation, the visual assessment by dispersion diagrams is useful. For greater precision, however, the other methods are needed. If the number of occurrences is large, i.e. with large samples, the use of the standard error of the difference is as sound as any other, and is often preferable in such cases. With moderate to small samples, Student's *t* Test should be applied, however, while in all cases it is desirable to use the 'best estimate' of the standard deviation, as it is sample values which are being compared.

In all the examples which have so far been analysed, however, the two sample means have always each been based on the same size of sample. This is not invariably so with data that the geographer may need to examine. It might be necessary, for example, to assess the relative importance of two different routeways as outlets for a rather inaccessible mining area. Sample traffic surveys are taken of the number of mining lorries using each of these two routes, and the mean values obtained are 150 lorries per week for Route A and 200 lorries per week for Route B. On such data, traffic flow diagrams have often been prepared, and a difference as great as this would probably be accepted at face value. Two important pieces of information are not provided in the above figures, however. One is an index of scatter, i.e. deviation, and the other is the number of weeks during which counts were made, i.e. the size of the sample. With these added the relevant data are as follows:

Route A—$\bar{x}_1 = 150$; $\hat{\sigma}_1 = 59{\cdot}9$; $n_1 = 20$
Route B—$\bar{x}_2 = 200$; $\hat{\sigma}_2 = 66{\cdot}5$; $n_2 = 10$

These values would obtain if the observed values were:

Route A: 40, 60, 80, 90, 100, 110, 120, 130, 140, 150, 150, 160, 170, 180, 190, 200, 210, 220, 240, 260.
Route B: 80, 110, 160, 180, 200, 200, 220, 240, 290, 320.

It can now be seen that on both routes numbers of lorries per week were highly variable. Moreover, a greater number of counts was made on one route than on the other, i.e. the sizes of the two samples differ. Such limitations and qualifications as these are often operative in this kind of study, and it is therefore necessary to apply fairly stringent tests before accepting the apparently marked difference between the routes as applying over the long term. The null hypothesis must first be posed:

H_0 = there is no difference between the two routes in terms of lorry frequency. Thus the inverse hypothesis is:
H_1 = a difference in lorry frequency between the two routes does exist. Then the t test must be applied:

$$t = \frac{|150 - 200|}{\sqrt{\frac{59 \cdot 9^2}{20} + \frac{66 \cdot 5^2}{10}}} = \frac{50}{\sqrt{\frac{3588}{20} + \frac{4422}{10}}} = \frac{50}{\sqrt{179 \cdot 4 + 442 \cdot 2}}$$

$$= \frac{50}{\sqrt{623 \cdot 9}} = \frac{50}{24 \cdot 93} = 2 \cdot 0 \text{ (approx.)}$$

$$\text{d.f.} = n_1 + n_2 - 2 = 20 + 10 - 2 = 28$$

On reading, $t = 2{\cdot}0$ against 28 degrees of freedom in Fig. 30 and taking $\alpha = 0{\cdot}05$ as the critical limit, it is found that this sample difference of 50 lorries per week, apparently so clear-cut, is *not* a significant one, for the value falls in the acceptance zone for the null hypothesis. Once again it must be stressed that this simply means that a significant difference has not been proven, not that it does not exist. It also indicates that more data are required, and with an increase in the number of traffic-flow surveys, especially on Route B, this question of degree of significance should be clarified one way or another. As was suggested in p. 121, the relative sizes of the samples should be proportional to the standard deviations.

Comparison of Coefficients of Variation

Finally, before moving to other methods of comparing differing sets of data in the succeeding chapters, there is yet a further modification of one of these methods which is of value in many geographical studies. Especially in climatology, though at times in other branches of the subject too, maps are prepared based on the *coefficient of variation* (V)—p. 42—and then comparisons are made between places with differing values of this coefficient. Far too seldom, however, has an investigation first been made to see whether the difference being explained or used is really significant statistically, or whether it is likely to be simply a chance occurrence.

Such an assessment can be made by a modification of the formula for the standard error of the difference between sample means. The latter relies on the standard error of the mean $\frac{\hat{\sigma}}{\sqrt{n}}$. As has been indicated earlier (p. 94) there is also a standard error of the standard deviation—$\frac{\hat{\sigma}}{\sqrt{2n}}$—and from this is derived the standard error of the coefficient of variation (V), i.e. $\frac{\hat{V}}{\sqrt{2n}}$. This latter value can readily be substituted for $\frac{\hat{\sigma}}{\sqrt{n}}$ in the formula for the standard error of the difference between means, to give a method of calculating the standard error of the difference between coefficients of variation. So instead of

$$\text{S.E.}\,|\bar{x}_1 - \bar{x}_2| = \sqrt{\frac{\hat{\sigma}_1^2}{n_1} + \frac{\hat{\sigma}_2^2}{n_2}} \quad \text{can be written}$$

$$\text{S.E.}\,|\hat{V}_1 - \hat{V}_2| = \sqrt{\frac{\hat{V}_1^2}{2n_1} + \frac{\hat{V}_2^2}{2n_2}}$$

A comparison may then be made between, for example, two rainfall stations, each with a 30-year record, but in one case with $\hat{V} = 13\%$ and in the other case $\hat{V} = 10\%$

$$\text{Thus: S.E.}\,|13 - 10| = \sqrt{\frac{13^2}{60} + \frac{10^2}{60}} = \sqrt{\frac{169}{60} + \frac{100}{60}} = \sqrt{\frac{269}{60}}$$

$$= \sqrt{4{\cdot}5} = 2{\cdot}1$$

Twice the standard error is 4·2 and the actual difference only 3·0, so that it is far from being a significant difference. Even if the value for $\hat{V}_1$ were increased to 14%, and the difference between $\hat{V}_1$ and $\hat{V}_2$ thus raised to 4%, this difference would still not be even probably significant.

It is thus both salutary and profitable to ensure that the differences between the sample mean values, or possibly between the sample variability values, under consideration possess a certain element of statistical significance. This equally applies to probability assessments, which are also based on sample data; the calculation of individual standard errors is outlined in Chapter 6, and these can be used for testing comparisons via the t test in a similar way to the examples above. At times this may mean that judgment must be deferred, or even that the absence of a statistically significant difference must be accepted. At other times a difference of sufficient significance is established for valid conclusions and further deductions to be based firmly and soundly upon it. It is in this role of focusing attention upon the legitimate cases, and indicating the degree of reliability or unreliability of marginal cases, that the methods outlined in this chapter can provide the greatest assistance to the geographer. Other methods, for dealing with more complex data, or with data in different forms, also exist, and some of these will be considered in the following two chapters.

CHAPTER 9

THE COMPARISON OF SAMPLE VALUES—II

(*The analysis of variance*)

In Chapter 8 several methods were presented by which it is possible to make some objective assessment of the validity of the differences between *two* sample mean values. Although problems of this sort are relatively frequent in the geographical field, there are also many occasions when a comparison is required between *more than* two sets of data. Such an assessment of whether the difference between several sets of data is significant or not is essential before any consideration is given to *what* is causing the difference. The sort of questions and problems that such a consideration may pose, and the methods by which they can often be resolved, are best approached through a specific example, and such an approach is adopted in the following pages. After this initial consideration, it will then be possible to employ the same methods in the analysis of other problems.

The Allocation of the Variance

Suppose that a survey were being made of agriculture in some part of the country, and sampling techniques were being employed. A stratified random sample of farms was made with the aim of comparing crop yields between the strata. These strata were defined in terms of the character of the soil, there being three strata related to fen peat soils, soils developed on Keuper marls and those on boulder clay. In each stratum the random sample consisted of ten farms, and on considering the cereal yields for these farms it was found that the average yield of the farms on fen peat soil was 24·3 bushels per acre, that of those on Keuper marl was 22·2 bushels per acre, while on boulder clay it was 21·0 bushels per acre. The problem thus presented is whether the difference between these three samples is such that it would be legitimate to claim that average cereal yields in that area vary significantly in relation to the parent material of the soil.

The first requirement is to set out the full sample data on which this comparison must be based (Stage I). These are tabulated below in the three strata outlined above, and it is seen that values vary markedly within each stratum as well as between the strata. Furthermore, if all the thirty values are grouped together the combined set of data will possess a variance (s^2) which reflects the tendency for the individual farm values to vary around an overall mean value. This being so, *any* stratification of the thirty farms into three groups, even if such stratification be done purely at random, is likely to lead to some difference between the means of the three strata. The question is therefore whether the difference between the strata used (which were based on a particular characteristic, i.e. soils) merely reflects such a random difference between any three groups of ten items each out of

Stage I. *Cereal yields of ten sample farms on the following soils*

	Fen peat	Keuper marl	Boulder clay
	24	17	19
	27	25	18
	21	24	22
	22	19	24
	26	28	23
	19	21	18
	25	20	21
	29	25	19
	26	19	25
	24	24	21
Average	24·3	22·2	21·0

these thirty items, or whether this observed difference is significantly greater than such a random difference would be. In other words, is the difference *between* the samples (referred to as the 'between sample difference') significantly greater than the differences that can be observed *within* each sample (referred to as the 'within sample difference')? If it is *not* significantly greater, then it could well be that the observed differences between the strata are only the result of chance grouping, in which case no proof of the influence of soils on crop yields can be obtained from this evidence. On the other hand, if the 'between sample difference' were significantly greater than the 'within sample difference' then it would be legitimate to assume that soil differences (as defined in the stratification) do lead to differences in crop yields. This is not to deny, of course, that many other factors will also affect crop yields, e.g. the quality of farm management or the amount of capital invested in the farm, for it is these other factors which lead to the observed differences between the sample items in each stratum, i.e. they comprise the 'within sample difference'.

The first necessity is therefore to divide the variance of the total set of thirty values into these two component groups, i.e. to allocate the amount of the overall variance that is due to 'between sample differences' and that part that is due to 'within sample differences'. It will be remembered that the variance is simply the average of the sum of the squares of the deviations from the average (p. 23), but to calculate this in detail can involve considerable labour. One expedient is to adopt an 'assumed average' as was done in the case of the short-cut calculation of the average and standard deviation in Chapter 3. Once again the accuracy of the assumed average does not affect the method of working; its sole effect is that the nearer the assumed average is to the actual average the smaller will be the numbers with which it is necessary to work.

In the present example this assumed average can conveniently be taken as 22, the choice being made largely by visual assessment—experience will lead to a sound choice being made in most cases. Once this assumed average has been chosen, the original data can be retabulated simply as differences from this value. This has been done in the following table (Stage II).

Stage II.

	ITEMS LESS 22				SQUARES OF 'ITEMS LESS 22'		
	2	−5	−3		4	25	9
	5	3	−4		25	9	16
	−1	2	0		1	4	0
	0	−3	2		0	9	4
	4	6	1		16	36	1
	−3	−1	−4		9	1	16
	3	−2	−1		9	4	1
	7	3	−3		49	9	9
	4	−3	3		16	9	9
	2	2	−1		4	4	1
Total	23	2	−10	Total	133	110	66

The values on the left represent the deviations of the individual items from the assumed average. To calculate the variance these deviations must be squared, and this has been done on the right-hand side in the table above under the heading 'Squares of "Items less 22" '. Having prepared these two tables it is convenient to add up each column so that the sum of each column is available for later calculations.

To obtain the 'overall variance', which must then later be allocated to 'between' and 'within' sample differences, these individual squares of the deviations from the mean must be summed. This is simply done by adding together the totals for the three columns of the samples, i.e. in this case by

$$133 + 110 + 66 = 309$$

However, the deviations which were squared in this connection were deviations from an *assumed* mean, and there is therefore the need to apply a 'correction factor' to allow for this. To calculate this 'correction factor' it is necessary to return to the tabulated 'Items less 22'. The totals of the three samples represent the differences which are left over because of the difference between the assumed average and the actual average. This set of differences has been incorporated in the overall variance of 309 already obtained above. If, therefore, the amount of variance that this contributes towards the 309 total can be obtained, a suitable correction can be applied. This can be done by calculating the variance of these sample totals, i.e. the totals of the sample columns are summed to get the total difference from the assumed average; this value is squared; and then this is divided by the number of items in *all* the samples

i.e. $23 + 2 + (-10) = 15$
$15 \times 15 = 225 = T^2$ (total differences squared)
$\frac{225}{30} = 7{\cdot}5 = \frac{T^2}{N}$ (N = total number of items)

The resultant value of $\frac{T^2}{N}$, i.e. 7·5 in this case, is the contribution to the overall variance value of 309 which is made by the difference between the assumed mean and the actual mean. Thus this value is the necessary 'correction factor' by which the total of the sum of the squares of the deviations from the assumed average (309) must be adjusted to get the sum of the squares from the *actual* average, i.e. $309 - 7{\cdot}5 = 301{\cdot}5$ = the sum of the squares. Finally, to obtain the variance itself, the sum of the squares must be divided by the number of occurrences. As these data are all samples, however, an element of safety is introduced by using the 'degrees of freedom' as in Student's t Test, i.e. degrees of freedom = N − 1 = 30 − 1 = 29. Dividing this number into the sum of the squares (301·5) would give the overall variance from the actual mean (10·40 approximately), but in the calculation there is no need to obtain the variance itself. It is the component elements of the variance, i.e. the 'sum of the squares of the deviations' and the 'degrees of freedom', that are required. Then, instead of allocating the variance to 'between' and 'within' sample differences, these two components of the variance can be allocated instead.

At this point, however, it is as well to summarize the calculations that have so far been made, and to set them out in a systematic and orderly manner. (Stage III). After tabulating 'Items less 22' and 'Squares of "Items less 22" ', and obtaining the totals of each column (p. 154), the following procedure should be adopted to obtain the components of the overall variance.

Stage III.

ITEMS LESS 22

Total 23	2	−10

T (sum of sample totals) = 15
N (total items) = 30

$$\text{Correction factor} = \frac{T^2}{N} = \frac{15^2}{30} = \frac{225}{30} = 7{\cdot}5$$

SQUARES OF 'ITEMS LESS 22'

Total 133	110	66

Total sum of the squares = sum of sample totals − correction factor = 309 − 7·5 = 301·5

Degrees of freedom = $N - 1$ = 30 − 1 = 29

Having thus obtained the overall picture the problem is to allocate the sum of the squares and the degrees of freedom to 'between sample' and 'within sample' groups. This is most easily done if those parts of these values that are due to 'between sample' conditions are first allocated. In any of the samples, if the overall average is applied to that sample then the total value for the sample would be that average multiplied by the number of items, i.e. in the present case, 22 × 10 = 220. In the first sample, however, the total differs from this by 23, while the second and third samples differ from it by 2 and 10 respectively. As it is the *between* sample value that is being assessed, it can be assumed that this overall deviation for any one sample is evenly distributed between each of the occurrences within that sample. In this way the *within* sample variation is eliminated.

Therefore, whereas the sum of the squares of the differences in any body of data would be $\Sigma\,(x - \bar{x})^2$, if the $(x - \bar{x})$ value is the same for each item in the sample, as postulated above, then $\Sigma\,(x - \bar{x})^2 = n(x - \bar{x})^2$. The following algebraic modification can then be made:

$$n(x - \bar{x})^2 = (\sqrt{n}.(x - \bar{x}))^2 = \left(\frac{n(x - \bar{x})}{\sqrt{n}}\right)^2 = \frac{(n\,(x - \bar{x}))^2}{n}$$

$$= \frac{(\Sigma\,(x - \bar{x}))^2}{n}$$

The value $\Sigma\,(x - \bar{x})$ is the total value of the deviations from the average in the sample concerned, i.e. 23, 2 and 10 in the present example. Thus the sum of the squares for the sample, with *within* sample differences eliminated, can be obtained by squaring this total deviation of the sample and dividing by the number of items in that sample. From this it follows that the total value of the sum of the squares resulting from *between* sample differences is obtained by summing these values for all the samples, i.e. in the present example, by

$$\frac{23^2}{10} + \frac{2^2}{10} + \frac{10^2}{10} = \frac{529}{10} + \frac{4}{10} + \frac{100}{10} = \frac{633}{10} = 63{\cdot}3$$

As in the present case the size of the sample is the same in all three instances, it is easier to sum the squares of the differences first and then divide this value by the number of items in each sample. The answer obtained by either of these methods is based on differences from the assumed mean, and therefore the correction factor must be subtracted from it to obtain the true answer. Thus the 'between sample sum of the squares' is $63{\cdot}3 - 7{\cdot}5 = 55{\cdot}8$. Having obtained this value, the 'within sample sum of the squares' is simply the amount that is left when this value is subtracted from the overall sum of the squares, i.e. $301{\cdot}5 - 55{\cdot}8 = 245{\cdot}7$.

The degrees of freedom can be allocated in the same sort of way. In the case of 'between sample' conditions the degrees of freedom are simply 1 less than the number of samples being considered. Here there are three such samples and so the 'between sample degrees of freedom' are $3 - 1 = 2$. Again, with this value obtained the 'within sample degrees of freedom' are simply obtained by calculating the difference between this value and the overall value, i.e. $29 - 2 = 27$.

Once more, however, it is as well to summarize this reasoning and these calculations and to set them out carefully as follows:

Stage IV.

'Between sample' conditions
sum of the squares:

$= \frac{1}{n}(a^2 + b^2 + c^2) -$ correction factor

(where n = no. of items per sample and a, b, c are the sample totals of differences)

$= \frac{23^2 + 2^2 + 10^2}{10} - 7{\cdot}5$

$= \frac{529 + 4 + 100}{10} - 7{\cdot}5$

$= \frac{633}{10} - 7{\cdot}5 = 63{\cdot}3 - 7{\cdot}5 = 55{\cdot}8$

degrees of freedom:
= no. of samples − 1
= 3 − 1 = 2

'Within sample' conditions
sum of the squares:
= total sum of squares − 'between sample' sum of squares
= 301·5 − 55·8 = 245·7

degrees of freedom:
= total degrees of freedom − 'between sample' degrees of freedom
= 29 − 2 = 27

Snedecor's Variance Ratio Test

In this way the overall sum of the squares and degrees of freedom (and thus the overall variance) have been allocated to the two categories as regards origin. Part of the overall variance was produced by differences *between* the samples, this amount of variance being obtained by dividing the appropriate sum of the squares by the equivalent degrees of freedom, i.e. 55·8/2 = 27·9. As these are samples only, this is known as the '*variance estimate*'. Equally the other part of the overall variance was produced by differences *within* the samples, the division of sum of the squares by degrees of freedom here being 245·7/27 = 9·1 = 'variance estimate'. Again, tabulation allows these features to be seen more clearly.

Stage V.

Source of variance (*a*)	Sum of squares (*b*)	Degrees of freedom (*c*)	Variance estimate (*b*/*c*)
(i) between sample	55·8	2	27·9
(ii) within sample	245·7	27	9·1

These estimates of the variance of 'between sample' and 'within sample' conditions must now be compared. The purpose of such a comparison is to see whether these variance estimates are so much alike that the differences between the samples simply reflect the differences within the samples, i.e. that no significant difference between the sample means can be assumed; or whether they are sufficiently dissimilar for a significant difference between the samples to be accepted. As this involves statistical testing, we begin by postulating a null hypothesis, as with the t test (pp. 142-144). By this is meant that one assumes that *no* significant difference exists between the samples, and that therefore the two variance estimates are not significantly different, and then tests to see the probability that such an assumption is justified. A direct comparison of the two variance estimates of 27·9 and 9·1 is not possible, however, because of the markedly different number of occurrences on which they are based, i.e. the 'degrees of freedom' are different in the two cases. To overcome this difficulty a test known as '*Snedecor's Variance Ratio Test*' is applied. This consists of a simple ratio which gives a value called 'Snedecor's F' which is then referred to tables; for a variety of critical probability limits, or Type I error levels, these indicate whether the null hypothesis should be accepted or rejected. This ratio is calculated as follows:

$$\text{Snedecor's } F = \frac{\text{greater variance estimate}}{\text{lesser variance estimate}} = \frac{27 \cdot 9}{9 \cdot 1}$$

$$F = 3 \cdot 07$$

The tables to which this value is referred are known as the 'percentage points of the F-distribution' and these are needed *at least* for the 5% and 1% levels ($2\frac{1}{2}$% and 0·1% are also useful at times). Summary versions of these tables are given in Table XXII. Thus if the F value falls into the 5% probability range there is no more than a 5% chance of error when one rejects the null hypothesis and accepts the inverse hypothesis. In other words, a difference of this order, in which the F value falls in the 5% probability range, is one that is *probably significant*. On the other hand, if the F value falls in the 1% probability range it means that a difference of the order of the observed one will occur by random grouping of the data into three ten-item samples only once in a hundred times, i.e. there is only a 1% probability that one would be wrong to accept that real differences exist between these sets of data.

To return to the example under study, it was seen that $F = 3{\cdot}07$. It is then necessary to locate this value on the appropriate part of

Table XXII

Percentage points of the F-distribution

$$\left(F = \frac{\text{greater variance estimate}}{\text{lesser variance estimate}}\right)$$

5% Level of Variance Ratio

Number of degrees of freedom of greater variance estimate

Number of degrees of freedom of lesser variance estimate	1	2	3	4	5	10	20	∞
1	161	200	216	225	230	242	248	254
2	18·5	19	19·2	19·2	19·3	19·4	19·4	19·5
3	10·1	9·6	9·3	9·1	9·0	8·8	8·7	8·5
4	7·7	6·9	6·6	6·4	6·3	6·0	5·8	5·6
5	6·6	5·8	5·4	5·2	5·0	4·7	4·6	4·4
10	5·0	4·1	3·7	3·5	3·3	3·0	2·8	2·5
20	4·3	3·5	3·1	2·9	2·7	2·3	2·1	1·8
∞	3·8	3·0	2·6	2·4	2·2	1·8	1·6	1·0

1% Level of Variance Ratio

Number of degrees of freedom of greater variance estimate

Number of degrees of freedom of lesser variance estimate	1	2	3	4	5	10	20	∞
1	4,052	5,000	5,403	5,625	5,764	6,056	6,192	6,366
2	98	99	99	99	99	99	99	99
3	34	31	29	29	28	27	27	26
4	21	18	17	16	16	15	14	13
5	16	13	12	11	11	10	9·6	9
10	10	7·6	6·6	6·0	5·6	4·8	4·4	3·9
20	8·1	5·8	4·9	4·4	4·1	3·4	2·9	2·4
∞	6·6	4·6	3·8	3·3	3·0	2·3	1·9	1·0

For full details see: D. V. Lindley and J. C. P. Miller, *Cambridge Elementary Statistical Tables*, Cambridge, 1953 (Table 7).

Table XXII. If the part for the 1% level is first considered it will be seen that at the top, from left to right, are set out the number of degrees of freedom of the *greater variance estimate*, while on the left, from top to bottom, are the number of degrees of freedom of the *lesser variance estimate*. By referring to p. 151 it will be seen that in the present case the former is 2 while the latter is 27. By setting the latter between 20 and infinity it is seen that the appropriate *F* number should be within the limits 4·6 to 5·8. On the 5% level part of Table XXII the same reference to these degrees of freedom shows that the appropriate *F* number should be within 3·0 and 3·5. As the actual value for the present example was 3·07, it can be said that the differences with which the problem was concerned are certainly not significant at the 1% level, when the null hypothesis would have to be accepted. In fact, by reference to more detailed tables it will be seen that the appropriate *F* value at the 5% level is 3·35, so that in this situation also, acceptance of the null hypothesis is necessary. The differences in cereal yields between these three samples of ten farms on fen peat (24·3 bushels per acre), ten on marl (22·2 bushels per acre) and ten on clay (21·0 bushels per acre) have therefore *not* been shown to be significant; it would be unjustified to assume a soil/cereal yield relationship for the area concerned—at least until further evidence proved otherwise.

Alternative Method

The method outlined above, based on an assumed mean and so needing a correction factor, can be replaced by a somewhat simpler method if calculators are available. In this case, the 'sum of the squares' can be directly computed from the original data given on p.153 (Stage I), by means of the following sub-totals:

Stage II.

(X_1) Fen peat	(X_2) Keuper marl	(X_3) Boulder clay
$\Sigma X_1 = 243$	$\Sigma X_2 = 222$	$\Sigma X_3 = 210$
$\Sigma X_1^2 = 5{,}985$	$\Sigma X_2^2 = 5{,}038$	$\Sigma X_3^2 = 4{,}466$

Stage III.

$$\text{Total sum of the squares} = \Sigma X^2 - \frac{(\Sigma X)^2}{N} \text{ (see p. 31)}$$

$$= 15{,}489 - \frac{675^2}{30}$$

$$= 301{\cdot}5$$

$$(\Sigma X = \Sigma X_1 + \Sigma X_2 + \Sigma X_3 = 243 + 222 + 210 = 675)$$

$$(\Sigma X^2 = \Sigma X_1^2 + \Sigma X_2^2 + \Sigma X_3^2 = 5{,}985 + 5{,}038 + 4{,}466 = 15{,}489$$

Total degrees of freedom $= N - 1 = 30–1 = 29$

Thus, in Stage III, the formula for the sum of the squares is as given on p. 31. It should be noted, however, that this employs the original values, not the difference from an assumed average. Therefore $\frac{(\Sigma X)^2}{N}$ is part of the basic formula, and should not be confused with the correction factor $\frac{(T^2)}{N}$ from the earlier formula.

'Between sample' conditions are obtained in Stage IV, where a similar formula for the sum of the squares is applied to the totals of the individual samples, and the answer divided by the number of items in each sample (as in the earlier form of computation). As can be seen, exactly the same answers are obtained, and the choice between these methods really depends on the facilities available.

Stage IV.

(i) '*Between sample*' *conditions*

$$\textit{sum of the squares} = \left(\frac{(\Sigma X_1)^2}{n_1} + \frac{(\Sigma X_2)^2}{n_2} + \frac{(\Sigma X_3)^2}{n_3}\right) - \frac{(\Sigma X)^2}{N}$$

$$= \left(\frac{243^2}{10} + \frac{222^2}{10} + \frac{210^2}{10}\right) - \frac{675^2}{30}$$

$$= 15{,}243{\cdot}3 - 15{,}187{\cdot}5$$

$$= 55{\cdot}8$$

degrees of freedom $= y - 1 = 3 - 1 = 2$ (when y = no. of sample totals)

(ii) '*Within sample*' *conditions*

sum of the squares = total sum of squares—between sample sum of squares

$$= 301{\cdot}5 - 558 = 245{\cdot}7$$

degrees of freedom = total degrees of freedom—between sample degrees of freedom

$$= 29 - 2 = 27$$

Samples of Different Sizes

As in the example so far used all samples were the same size, it may be useful to consider a case where this does not apply. Suppose that in a rather underdeveloped part of the world it is decided to try to assess by sampling methods the population per village. In the area under study it is known that four different tribal groups exist, and it is suspected that the size of the village unit may vary with the tribe. In the absence of data on variance, the sample is stratified proportionately to the estimated average number of people in villages occupied by these four different tribes, with the results set out overleaf. As can be seen, not only does the average population per village vary with the tribe, but also the size of the sample differs from one tribe to the other. The question is therefore whether or not the tribal groupings significantly affect the average population per village.

	Tribe A	Tribe B	Tribe C	Tribe D
	150	150	300	350
	550	500	550	800
	250	400	400	300
	150	350	250	350
		250	350	700
		450	550	550
				450
				500
Average population per village	275	350	400	500

The analysis is carried out in the same way as before, except that the difference in the size of the samples must be remembered especially when calculating the 'between sample' sum of the squares. If the 'assumed mean' method is used, the assumed mean can in this case be 400, and the values retabulated as follows:

Items less 400					Squares of 'items less 400'			
A	B	C	D		A	B	C	D
−250	−250	−100	−50		62,500	62,500	10,000	2,500
150	100	150	400		22,500	10,000	22,500	160,000
−150	0	0	−100		22,500	0	0	10,000
−250	−50	−150	−50		62,500	2,500	22,500	2,500
	−150	−50	300			22,500	2,500	90,000
	50	150	150			2,500	22,500	22,500
			50					2,500
			100					10,000
Total −500	−300	0	800	Total	170,000	100,000	80,000	300,000

$T = 800 - 800 = 0$

$N = 4 + 6 + 6 + 8 = 24$

$\text{C.F.} = \frac{0^2}{24} = 0$ (i.e. the actual mean is also 400)

'Between sample' sum of squares =

$$\frac{(-500)^2}{4} + \frac{(-300)^2}{6} + \frac{(0)^2}{6} + \frac{(800)^2}{8} - 0$$

$= 62{,}500 + 15{,}000 + 0 + 80{,}000 - 0$

$= 157{,}500$

degrees of freedom $= 4 - 1 = 3$

Total sum of squares $= 650{,}000 - 0 = 650{,}000$

Degrees of freedom $= 24 - 1 = 23$

'Within sample' sum of squares $= 650{,}000 - 157{,}500 = 492{,}500$

degrees of freedom $= 23 - 3 = 20$

Source of variance	Sum of squares	Degrees of freedom	Variance estimate
'Between sample'	157,500	3	52,500
'Within sample'	492,500	20	24,625

$$F = \frac{52{,}500}{24{,}625} = 2{\cdot}13$$

If the requisite degrees of freedom are read on the 1% and 5% tables (3 degrees for the greater variance estimate and 20 degrees for the lesser) it will be seen that the following F values are required

1% level $F = 4{\cdot}9$

5% level $F = 3{\cdot}1$

Observed value $F = 2{\cdot}13$

Thus the observed value falls on the 'accept null hypothesis' side of even the 5% line, so that little reliance can be placed on the significance of the differences between the four sample groups under study. In this case, therefore, despite what may appear to be quite large average differences between the tribal groups specified, in terms of the average population per village, these differences cannot legitimately be classed as even 'probably significant'. The different sizes of the samples obviously affect this answer to some extent, and an increase in the sampling fraction may well lead to a rather different answer. The analysis of variance presents a reasonably clear-cut verdict however, that on the basis of the sample data given above the difference *cannot* be classed as a significant one.

Thus the 'analysis of variance' provides a very valuable tool by which several sets of data may be compared, and an objective assessment be made of the significance of the apparent differences between them. Many more complex and refined analyses can also be carried out by modifications of this method, but recourse must be had to more advanced texts for examples of these. Moreover, it must be stressed that considerable care needs to be taken with this method at all times, for a large number of assumptions underlie its use. These can be checked from more advanced texts, and at research level it is essential that these assumptions are not ignored.

CHAPTER 10

THE COMPARISON OF SAMPLE VALUES—III

(*Non-parametric tests*)

In the two previous chapters methods have been presented by which it is possible to assess the statistical significance of differences between sample data, these differences being reflected in the sample mean values and also in the variances of these samples. Many cases occur, however, in which absolute values such as these may not be available although the frequency distribution of the data, based on some sort of grouping, can be obtained. It is therefore of value to consider methods by which the significance of such sample differences can be assessed. Moreover, all the tests of significance outlined earlier assume that the body of data fits, or approximates to, the normal distribution curve, i.e. they are parametric tests. When the body of data is markedly skew, however, it is often advantageous to effect a comparison in terms of the frequency distribution, rather than transforming the data and applying the previous tests. For these purposes, it is to the non-parametric tests that we must turn.

The Chi-squared Test

One method by which such comparisons may be made is known as the Chi-squared (χ^2) Test. This is a relatively easy test to apply, but it is *essential* that the data being considered are in the correct form and that the problem is a suitable one for this method. This value—χ^2—tests whether the *observed frequencies* of a given phenomenon differ significantly from the *frequencies* which might be *expected* according to some assumed hypothesis. This assumed hypothesis must be carefully defined and clearly thought out and understood, so that the results can be correctly interpreted. Furthermore, the data must be in the form of *frequencies* and NOT in absolute values.

The Calculation of χ^2

The most effective way to understand the characteristics and qualities of this method is to follow through several examples, so that the various possible difficulties are encountered and ways of circumventing them are seen in a specific context. At the same time the general approach can be outlined so that the method can be applied in other cases. It is desirable at this stage to concentrate solely on 'single sample' cases, leaving multiple samples or variables for later consideration. The first problem to be analysed in this way can be expressed as follows. Suppose that a study is being made of the number of farm sites in relation to the locational characteristics of those sites. Over an area of diversified relief a sample consisting of 200 farms is made, these farms being grouped into several categories depending on the physical character of the site—alluvium, terrace, steep slopes, limestone plateau and sandstone plateau. The values in each category are set out below.

Site	No. of farm sites	Types of terrain as % of all land
1. Alluvium	10	10
2. Terrace	100	35
3. Steep slopes	2	10
4. Limestone plateau	38	25
5. Sandstone plateau	50	20
Totals	200	100

Also in this table is the amount of land of each of these five types expressed as a percentage of all the land in the area under study. Clearly the distribution of farms between these different types of terrain is partially related to the amount of land of each type—thus terraces are the most frequent type of terrain and also contain the greatest number of farms, while the two types of terrain that occur least frequently also have the two smallest numbers of farms. On the other hand, the distribution of farms would also seem to partially indicate some preferential choice of site between these five possibilities—thus both the terrace and the sandstone plateau would seem to have a greater number of farms than their relative areas would suggest, while the other three sites are all under-represented to some degree. In trying to find an explanation for the spatial distribution of farm sites in such a situation, one problem which has to be solved is the relative importance of the two tendencies indicated above. If the number of farms on a given type of terrain is mainly a reflection of the frequency with which that type of terrain occurs then it cannot be argued that the characteristics and qualities of that type of terrain are factors influencing farm sites. Conversely, if the frequency with which farms occur on given terrains is *not* mainly a reflection of the frequency with which these terrains occur, then it would seem legitimate to argue that the different terrains possess characteristics and qualities which influence the siting of farms. Thus a causal relationship, possibly weak and possibly strong, may be established between types of terrain and the occurrence of farm sites.

To test which of these two possibilities is most likely it is necessary first to set up a 'null hypothesis', i.e. it is necessary to postulate that the observed distribution of farm sites could be reasonably expected in the light of the proportions of different types of land or, in other words, that there is *no* significant difference between these five groups (or terrains) as regards a preferential frequency of farm siting, the scattering of the farms simply representing a random distribution over the whole area. It is this null hypothesis (written in summary form as H_0) that is tested by χ^2, and which must be carefully and adequately posed.

This χ^2 value is calculated as follows. Let the 'observed' values (i.e. those that actually occur) be written as O, as set out below. Beneath these are written the values which would occur if the postulated null hypothesis really applied to the full. These are known as the 'expected' values, written as E.

	Group 1	Group 2	Group 3	Group 4	Group 5
'Observed' frequency	O_1	O_2	O_3	O_4	O_5
'Expected' frequency	E_1	E_2	E_3	E_4	E_5

From these the χ^2 value is obtained by the formula

$$\chi^2 = \sum \frac{(O - E)^2}{E}$$

$$\chi^2 = \frac{(O_1 - E_1)^2}{E_1} + \frac{(O_2 - E_2)^2}{E_2} + \frac{(O_3 - E_3)^2}{E_3} + \frac{(O_4 - E_4)^2}{E_4} + \frac{(O_5 - E_5)^2}{E_5}$$

Thus for each category the amount by which the observed frequency differs from the expected frequency is squared, and then related to the expected frequency itself. This is akin to the procedure for calculating the variance and then relating it (or the standard deviation) to the mean value from which actual conditions varied (see

p. 42). When these figures for each category are summed this gives the total sum of the squares of the difference between observed and expected values. Each difference is divided by the appropriate value because the value from which the observed data deviate is different in each group, in contrast with the deviations from the mean just mentioned. The division by E is therefore to eliminate this variable so that the summing of the separate squares becomes legitimate. Once the χ^2 value is obtained in this way, it can be referred to the appropriate table or graph and read off against the degrees of freedom. These, as in techniques discussed earlier, are obtained by subtracting 1 from the number of occurrences, i.e. the number of groups or categories. This table or graph will yield a value which gives the percentage probability of making a type I error (p. 143).

In the actual example which was commenced earlier, the observed values are available, while the expected values (based on the null hypothesis on p. 168) would be proportional to the amount of land in each category. Thus, as 10% of the land is alluvium it is to be expected, on the basis of the null hypothesis, that 20 out of 200 farms would be on alluvium. These expected values are given below with the observed values.

	1	2	3	4	5
O.	10	100	2	38	50
E.	20	70	20	50	40
$O - E =$	-10	30	-18	-12	10
$(O - E)^2 =$	100	900	324	144	100
$\frac{(O - E)^2}{E} =$	5·0	12·9	16·2	2·9	2·5

For each category it is a simple matter to calculate $(O - E)$, $(O - E)^2$ and $\frac{(O - E)^2}{E}$ as has been tabulated above. These can then be entered into the formula

$$\chi^2 = \sum \frac{(O - E)^2}{E}$$

i.e. $\chi^2 = 5{\cdot}0 + 12{\cdot}9 + 16{\cdot}2 + 2{\cdot}9 + 2{\cdot}5$
$= 39{\cdot}5$

This value reflects the sum of the squares of the deviations of observed conditions from the expected conditions.

As for the degrees of freedom, these are obtained by $N - 1 = 5 - 1 = 4$. By referring to the graph in Fig. 31 it will be seen that a χ^2 value of 39·5 with 4 degrees of freedom yields a probability value of less than 0·1%. This means that even if the critical value were $\alpha = 0.001$, the null hypothesis would still be rejected, and there would be less than one chance in a thousand of being wrong. There is therefore a more than 99·9% probability that the *inverse* of the

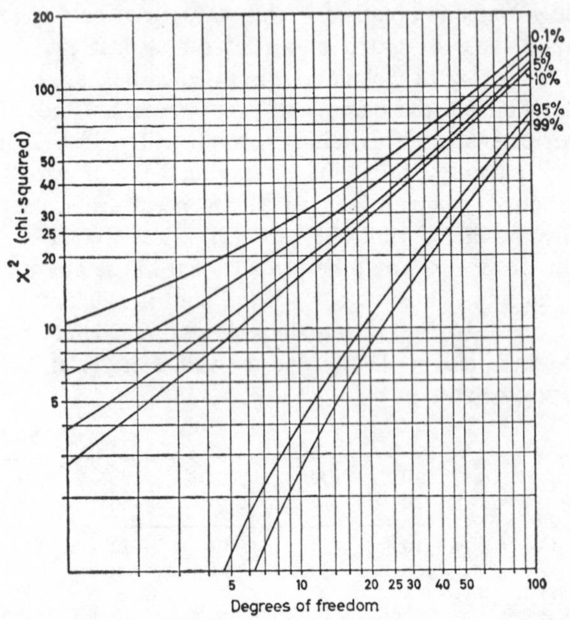

Figure 31. Graph for the Chi-squared Test (based on data in D. V. Lindley and J. C. P. Miller, *Cambridge Elementary Statistical Tables*, Table 5)

null hypothesis (written in summary form as H_1) is correct, i.e. that there is some preferential location of farm sites in relation to the various terrains. In other words, the percentage probability that farm sites are distributed between the different terrains in relation to the frequency with which these occur is very small indeed, and it would seem almost certain that the characteristics and qualities of the type of terrain *do* significantly affect the frequency with which farm sites occur.

A further illustration of the application of the χ^2 Test in this its simpler form may help to show more clearly both how the null hypothesis can be framed in different circumstances and also the various problems that can be analysed in this way. The first example illustrated the use of the test to assess whether or not there is a significant difference between observed frequencies and those to be expected according to some principle or theory—in this case, that the numbers of farm sites reflects the size of the area involved. It can also be used, however, to test whether or not a particular sample is likely to have been taken from a population of some specified form. For example, in a study of vegetation, it may have been established by prior work that a particular plant community consists of four major species in the occurrence ratio of 6 : 4 : 3 : 2. When another area is sampled, the same species are found to be most common, the number of occurrences of these plants in the sample being 130, 100, 40 and 30. These are seen to be in the same relative order as they are in the specified plant community, but their relative frequency is not exactly the same. The problem is then posed as to whether or not it is legitimate to include the sample data with the specified plant community.

The data can be tabulated as follows, with the 'expected' values consisting of the 300 sample items divided according to the ratios characteristic of the specified plant community, i.e. the 'population of some specified form' mentioned above. The null hypothesis (H_0) to be postulated is that there is *no* difference between the sample

	Plants				
	(1)	(2)	(3)	(4)	Total
Observed (sample area)	130	100	40	30	300
Expected (from the ratios for the community)	120	80	60	40	300

frequency and that based on the specified plant community, and this can be tested by the χ^2 Test.

$$\chi^2 = \frac{10^2}{120} + \frac{20^2}{80} + \frac{20^2}{60} + \frac{10^2}{40}$$

$$= 0{\cdot}8 + 5{\cdot}0 + 6{\cdot}7 + 2{\cdot}5 = 15{\cdot}0$$

Degrees of freedom $= N - 1 = 4 - 1 = 3$

On referring these values to the graph in Fig. 31 it will be seen that the null hypothesis can be rejected with less than a 1% chance of being wrong; therefore the inverse hypothesis (H_1), that there is a significant difference between the sample frequency and the specified population frequency, is probably the correct one. It would then not be valid to include the vegetation from which the sample was taken with the particular plant community with which we have been concerned, despite the occurrence of the same four major plants in the same relative order, so long as the *frequency* of occurrence is a critical criterion for such inclusion.

Yet further application of the single sample case of the χ^2 Test is worth consideration, for it enables an answer to be given to the question as to whether or not the sample that has been drawn is truly a random sample from the total population. For example, perhaps a study is to be made of the shopping facilities of a given town, where it is known that of an overall total of 100 shops, 50 of them are food shops of one sort or another, 30 of them sell consumer goods such as furniture, electrical appliances, etc., and the remaining 20 form a miscellaneous category. For reasons of time, and to enable a detailed study of each store to be made, a sample of only 30 shops is taken, of which 19 are food shops, 7 sell consumer goods and there are 4 in the miscellaneous group. Is this truly a random sample, from which generalizations concerning the total shopping facilities may be made, or is it too heavily biased in one direction?

The appropriate values can be tabulated as before, and a null

	Shop categories			
	Food	Consumer goods	Miscellaneous	Total
Observed (sample)	19	7	4	30
Expected (from the overall proportions)	15	9	6	30

hypothesis (H_0) postulated that there is no difference between the sample and total population distributions. To test this the χ^2 values are obtained:

$$\chi^2 = \frac{4^2}{15} + \frac{2^2}{9} + \frac{2^2}{6}$$

$$= 1{\cdot}07 + 0{\cdot}44 + 0{\cdot}67 = 2{\cdot}18$$

Degrees of freedom $= N - 1 = 3 - 1 = 2$

By reference to Fig. 31 it will be seen that even if $\alpha = 0{\cdot}10$ the null hypothesis must be accepted. A significant difference has not been established, and it is at least reasonable to *assume* that the sample and the total population are the same. It must be remembered, however, that it is never possible to *prove* this assumption with complete certainty.

Influence of Size of Sample

The results thus obtained are subject to the priviso that they are valid only in terms of the available evidence. The limitations imposed by this are the same as those resulting from the 'size of the sample' in examples in earlier chapters. In the α^2 Test, too, changes in the size of the sample can affect the resultant conclusions, even if the proportion of occurrences within any one category remain unaltered. Thus suppose that in a survey of the agriculture of an area attention was focused on the differences between dairying and non-dairying farms. It is known that, overall, 60% of the farms engage in dairy activities while 40% do not, but as the total number of farms is too large for detailed study a sample survey is carried out instead. Initially, a small sample of 20 farms is taken, of which 16 engage in dairying and only 4 do not; it is therefore desirable to assess whether or not this represents an adequately random sample. The null hypothesis (H_0) must be that there is no significant difference between the sample distribution and that of the total population, the validity of this hypothesis being assessed by the χ^2 test. The observed values are set out below, while the expected values on this basis are obtained by sharing the 20 farms in the ratio of 3 : 2 between dairying and non-dairying groups respectively.

	No. of farms				
	(O)	(E)	$(O-E)$	$(O-E)^2$	$\frac{(O-E)^2}{E}$
Dairying farms	16	12	4	16	1·33
Non-dairying farms	4	8	−4	16	2·0

$$\chi^2 = 1{\cdot}33 + 2{\cdot}0 = 3{\cdot}33$$

Degrees of freedom $= 2 - 1 = 1$

From Fig. 31 it will be seen that with these values for χ^2 and degrees of freedom, if $\alpha = 0{\cdot}05$ then the null hypothesis must be accepted, while if $\alpha = 0{\cdot}10$ it can be rejected On the other hand, the probability value is sufficiently close to the critical boundary of 5% for one to be loathe to base the results of the whole study upon such a sample without further consideration.

It may therefore be decided to increase the size of the sample by 50%, i.e. to make it consist of 30 farms. If the same proportion was found to comprise dairying as in the earlier smaller sample, this would mean that 24 dairying farms and 6 non-dairying would form the sample. These, and the expected values obtained by sharing the sample of 30 farms in a ratio of 3 : 2 again, are given below and the usual calculations made.

No. of farms

	(O)	(E)	$(O - E)$	$(O - E)^2$	$\frac{(O - E)^2}{E}$
Dairying farms	24	18	6	36	2
Non-dairying farms	6	12	−6	36	3

$$\chi^2 = 2 + 3 = 5$$

Degrees of freedom = 2 − 1 = 1

By referring these values, based on a larger sample, to Fig. 31, it will be seen that the probability of being in error when rejecting the null hypothesis has been reduced to 2·5%. It is therefore highly probable (97·5%) that the inverse of the null hypothesis is true, and therefore one cannot accept the sample as a truly random one. An element of bias has been introduced, perhaps unwittingly, and the sampling procedure may need revision. Thus the larger size of the sample, with the proportions remaining the same, clarifies the position. It must be realized, however, that in reality the new larger sample will almost certainly yield proportions of farms with and without dairying which differ from those of the first sample. They are more likely to be nearer the true proportions, however, because of the larger size of the sample.

The Comparison of Two or More Variables

So far in this consideration of the χ^2 Test only simple examples have been used, in which there has been in each case only *one* set of variable conditions—the frequency of farm sites upon different terrains; the frequency of different plants, different shops or different types of farming. In many problems, on the other hand, it may be desirable to compare *two or more* different sets of variable conditions, where, for example, two sets of data have different frequency distributions and it is necessary to ascertain whether these differences in frequency distributions are statistically significant or not.

A specific problem may help to clarify the matter. Suppose that a study is being made of the distribution of woodland over an area. It is seen that it is distributed very irregularly and the study is aimed at presenting some valid explanation of this irregular distribution. The woodland distribution is therefore compared with the distribution of various possible causative factors, amongst which is the form of land tenure. On consideration of the sample land holdings, which totalled 300 in all, it was seen that 90 of them were private estates while 210 were tenant farms. The obtaining of exact acreages and percentages of the holdings under woodland not proving possible for one reason or another, the holdings were instead grouped as to whether less than 10%, 10–20% or more than 20% of the holdings were under woodland. The values obtained are set out below, these representing the 'observed' frequencies to be used in the χ^2 Test. Furthermore, the differences between the frequency distributions for the two types of land holding can also be clearly seen. In attempting to assess whether or not these differences are statistically significant, it is again necessary to set up a null hypothesis, which in this case would be that there is no difference between private estates and tenant-farms in terms of the proportion of land under woodland.

	No. of holdings with given % of holding under woodland			
	>20%	10–20%	<10%	Totals
Private estates	30	45	15	90
Tenant farms	30	105	75	210
Totals	60	150	90	300

To obtain the 'expected' frequencies so that the χ^2 Test can be applied requires a certain amount of simple calculation plus an understanding of what is involved. It can most readily be explained by working in terms of both the above example and an idealized case at the same time. In each case it is necessary to remember that the totals for both the columns and the lines in the tabulation are fixed by the observed conditions, as was also true in the simpler examples considered earlier. The idealized case which will be used is set out below. The columns are shown by a, b, c etc. and their totals by x, y, z, while the lines are indicated by A, B, etc. and their totals by Y, Z. The overall total of items in the table is given by N.

	a	b	c	Totals
A	$\frac{Yx}{N}$			Y
B				Z
Totals	x	y	z	N

The first question to be asked is 'what is the probability of values occurring in Line A (or Private Estates)?' Clearly from the specific example it is $\frac{90}{300}$, i.e. the total of the line divided by the overall total number of occurrences. In the idealized case this would be $\frac{Y}{N}$. The second question is then 'what is the probability of values occurring in Column *a* (or >20% woodland)?' Equally obviously from the specific example this is $\frac{60}{300}$, i.e. the total of the column divided by the overall total number of occurrences. In the idealized case this would be $\frac{x}{N}$. From these two questions arises the third, i.e. 'what is the probability of values occurring in *both* Line *A and* Column *a* at the same time, i.e. of falling in Square *Aa*?' This is obtained by multiplying together the two probabilities set out above, by applying the 'Multiplication Law' mentioned on p. 72. Thus, in the case of the specific example, the probability of a holding falling into the category of a private estate with more than 20% of the land under woodland would be

$$\frac{90}{300} \times \frac{60}{300} = \frac{5{,}400}{90{,}000} = \frac{54}{900} = 0{\cdot}06$$

In the case of the idealized example this would be

$$\frac{Y}{N} \cdot \frac{x}{N} = \frac{Yx}{N^2}$$

If these values $\left(0{\cdot}06 \text{ and } \frac{Yx}{N^2}\right)$ give the *probability* of an occurrence falling in Square *Aa*, then the actual *frequency* or number of occurrences can be obtained by multiplying this probability by the overall total of occurrences:

i.e. in the specific case $\frac{6}{100} \times 300 = 18$

in the idealized case $\frac{Yx}{N^2} \times N = \frac{Yx}{N}$

In other words, the *expected frequency* in any given square is simply obtained by multiplying the appropriate Column and Line Totals, and dividing this by the Overall Total of Occurrences. Thus in the idealized example the expected frequency in Square *Bc* would be $\frac{Zz}{N}$ and in Square *Ac*, $\frac{Yz}{N}$.

From this very necessary digression to explain the mechanism by which the expected values can be calculated, it is now time to return to the calculations in the specific example introduced above. The expected frequencies for this example are calculated as follows:

	% of holding under woodland			
	>20%	10–20%	<10%	Totals
Private estates	$\frac{90 \times 60}{300}$	$\frac{90 \times 150}{300}$	$\frac{90 \times 90}{300}$	90
Tenant farms	$\frac{210 \times 60}{300}$	$\frac{210 \times 150}{300}$	$\frac{210 \times 90}{300}$	210
Totals	60	150	90	300
Private estates	18	45	27	90
Tenant farms	42	105	63	210
Totals	60	150	90	300

With both observed (p. 175) and expected values thus obtained it is possible to calculate χ^2 by the same formula as before, i.e. $\sum\frac{(O-E)^2}{E}$

$$\text{Thus } \chi^2 = \frac{(30-18)^2}{18} + \frac{(45-45)^2}{45} + \frac{(15-27)^2}{27} + \frac{(30-42)^2}{42} + \frac{(105-105)^2}{105} + \frac{(75-63)^2}{63}$$

$$= \frac{144}{18} + \frac{0}{45} + \frac{144}{27} + \frac{144}{42} + \frac{0}{45} + \frac{144}{63}$$

$$= 8{\cdot}0 + 5{\cdot}33 + 3{\cdot}43 + 2{\cdot}29$$

$$= 19{\cdot}05$$

When referring this value to the χ^2 Graph the degrees of freedom are also required. These must be obtained both for the columns *and* the lines, employing the usual method of $(n - 1)$ *in each case*. Thus if n_1 is the number of lines then the degrees of freedom will be $(n_1 - 1)$. Equally, if n_2 is the number of columns then the degrees of freedom will be $(n_2 - 1)$, while the overall degrees of freedom will be the product of these two values, i.e. $(n_1 - 1)(n_2 - 1)$. In the present case these two values are $2 - 1 = 1$ and $3 - 1 = 2$ respectively, so that the product, i.e. the overall degrees of freedom, is $2 \times 1 = 2$. The reader may check this value by inserting any value into two of the six squares ensuring that they are not both in the same column. With the totals remaining constant it will be found that the other four values must automatically follow. If now a χ^2 value of 19·05 is read off against 2 degrees of freedom on Fig. 31 it will be seen that the relevant probability value is *less than* 0·1%. This means that the probability of wrongly rejecting the null hypothesis is less than 0·1%. The null hypothesis is thus not tenable; rather it can be said that there is a highly significant difference between private estates and tenant farms in terms of the proportion of their holding that is likely to be under woodland.

In considering this example the thought may well have occurred that if the actual amount of woodland on each holding had been known, then the tests of significance outlined in the previous two chapters could have been applied. While this is true, the χ^2 Test has several points to recommend it. First, the data may only be available in a grouped form and actual values not known, as was specified in this example at the outset. Second, such tests as Student's t make assessments on the basis of two parameters—arithmetic average and standard deviation. χ^2, in contrast, compares the whole frequency distribution, especially if a large number of groups are used. This is very important when the frequency distribution of the sets of data being compared are not normal, for all formulae based on averages and standard deviations assume a normal or near-normal distribution curve. In the present example, the data for private estates is negatively skew, and that for tenant farms is positively skew, while the overall distribution is but slightly skew in a positive sense. For these reasons, the χ^2 Test is to be preferred in this and similar cases, such as those that are illustrated by the following two examples, which again involve the comparison of two sets of variables.

One sort of problem that can be studied in this way is represented by the following set of conditions. Across an upland area a study is being made of the depth of the peat layer and the angle of slope of the land where such depths are measured. A simple inspection suggests that peat is markedly deeper where the angle of slope is less than 5° than where it is more than 5°. Actual values of recordings under these two sets of slope conditions vary within certain limits, as is shown below where the observations are grouped according to peat depth.

	No. of borings with a given peat depth			
	>6 ft.	3–6 ft.	<3 ft.	Totals
<5° slope	10	20	6	36
>5° slope	3	30	21	54
Totals	13	50	27	90

Once again these three distributions are each skew, and not all in the same direction. Furthermore, the data are not in terms of absolute values but in groups or categories. When trying to test whether a significant difference does occur between slopes of less than, and more than, 5° the χ^2 Test can well be employed. The null hypothesis in this case is that there is *no* difference between conditions on the two different sets of slopes, and expected values can therefore be simply obtained on this basis by the method used in the last example. In other words, the expected value at any place is obtained by multiplying the 'line total' by the 'column total' and then dividing by the 'overall total'. The resultant values are set out below.

	Expected values of no. of borings with a given peat depth			
	>6 ft.	3–6 ft.	<3 ft.	Totals
<5° slope	5·2	20	10·8	36
>5° slope	7·8	30	16·2	54
Totals	13	50	27	90

The calculation of χ^2 thus becomes

$$\frac{4{\cdot}8^2}{5{\cdot}2} + \frac{4{\cdot}8^2}{10{\cdot}8} + \frac{4{\cdot}8^2}{7{\cdot}8} + \frac{4{\cdot}8^2}{16{\cdot}2} = 4{\cdot}43 + 2{\cdot}13 + 2{\cdot}95 + 1{\cdot}42 = 10{\cdot}93$$

As for the degrees of freedom, these are $(3 - 1)(2 - 1) = 2 \times 1 = 2$. By reference to Fig. 31 it can be seen that the null hypothesis can be safely rejected with no more than a 1% chance of being wrong, so it can be said that the observed values represent a significant difference in terms of peat accumulation between slopes of less than, and more than, 5°.

As a final example a different theme can be considered. In analysing the industrial character of two large towns, the question of the size of industrial establishment may be considered, and one way of doing this is to use the numbers of people employed by each firm. Frequently, however, it is not possible to obtain precise numbers in such studies, though it is usually possible to allocate each firm to a category which consists of a *range* of values. Thus in the present example it is possible to allocate firms to the following four groups—those that employ 2,000 or more people, 500 to 1,999 people, 100 to 499 people, and less than 100 people respectively. Having done this, with the observed values set out below, it is necessary to test whether any significant difference exists between these two towns in terms of the size of industrial firms.

	No. of firms employing given numbers of people				
	2,000+	500–1,999	100–499	<100	Totals
Town A	10	250	350	50	660
Town B	8	240	400	72	720
Totals	18	490	750	122	1,380

The null hypothesis in this case would be that there is *no* significant difference between the two towns, and the expected values can therefore be allocated in proportion to the various totals given above, as has been done in the last two examples. The expected values for these conditions are given below, and can be checked by the reader (for method see p. 176).

	Expected no. of firms employing given numbers of people				
	2,000+	500–1,999	100–499	<100	Totals
Town A	8·6	234	359	58·4	660
Town B	9·4	256	391	63·6	720
Totals	18	490	750	122	1,380

From these values χ^2 can be calculated, being in this case

$$0{\cdot}23 + 1{\cdot}09 + 0{\cdot}23 + 1{\cdot}21 + 0{\cdot}21 + 1{\cdot}00 + 0{\cdot}21 + 1{\cdot}11 = 5{\cdot}29$$

Degrees of freedom are here 3, by the same means of counting as in earlier examples. Reference to Fig. 31 indicates that the null hypothesis must be accepted, for rejection would involve a chance of error that is greater than 10%. This being so, it is unjustified to postulate a difference of any significance between these towns in terms of size of industrial firms.

It can thus be seen that many and varied problems of a geographical nature can be analysed by means of the χ^2 Test, which enables an objective assessment to be made. Furthermore it is often of great value in connection with data obtained from mapped distributions, where specific quantitative values are not available. Provided that the *frequency* with which conditions fall into specified categories can be established, then this test can be applied. The examples used have all been of a fairly simple and straightforward character, but more complex problems could be analysed by the use of this method. It will be found, however, that the working principles follow those outlined here. The more complex the problem, on the other hand, the more important it is that the terms of the null hypothesis be framed with care and the interpretation of results be done intelligently. In all cases, simple or complex, it must be remembered that the test is applied to *frequencies*, not to absolute values or proportions. One final point, which has been observed without comment in the above examples, is that, when degrees of freedom are more than 1, the χ^2 Test does not really work if the expected frequency in more than 20% of the cells is less than 5. If this is found to occur, then two or more cells must be grouped together until this expected value of 5 is obtained in at least 80% of the cells. Provided that the necessary care is taken, along the lines indicated above, this test can prove of considerable value in geographical work.

The Kolmogorov-Smirnov Test

Apart from the χ^2 Test, there are a large number of other non-parametric tests which can be used for the analysis of non-interval data, and it will prove useful to consider several applications of one of these. This is the Kolmogorov-Smirnov Test which, like the χ^2

Test, is applicable both to one sample and two sample problems; only *ordinal* data (p. 131) can be used, however.

One Sample Problems

As indicated in terms of χ^2 (p. 171), there is often the need to test a sample against some predetermined population of which the distribution characteristics are known. For example, one may be investigating the frequency of daily rainfall amounts between given limits, and for a long-term period it is known that these frequencies are as follows:

	Rainfall in a day (inches)			
	0″	0·01″–0·04″	0·05″–0·20″	>0·20″
% of days	50%	25%	20%	5%

For one particular 30 day month, however, the actual values were respectively 12, 6, 9, and 3 for these four categories. There is clearly some difference here from the overall distribution, but it is desirable to test the extent to which it differs to assess whether this difference is statistically significant or not. The null hypothesis (H_0) is therefore that no difference exists between the conditions of the particular month and those of the overall period.

To apply the Kolmogorov-Smirnov Test, the first need is to change both sets of values into proportions:

	Rainfall in a day (inches)			
	0″	0·01″–0·04″	0·05″–0·20″	>0·20″
Overall proportions	0·50	0·25	0·20	0·05
Observed proportions	0·40	0·20	0·30	0·10

Following this, these proportions must be accumulated, and then the difference between the values for the observed data and those for the overall data for each rainfall category must be obtained, as set out below.

	Rainfall in a day (inches)			
	0″	0·01″–0·04″	0·05″–0·20″	>0·20″
No. of items (total = 30)	12	6	9	3
Overall cumulative distribution	0·50	0·75	0·95	1·0
Observed cumulative distribution	0·40	0·60	0·90	1·0
Difference	0·10	0·15	0·05	0·0

It is with the *largest* of these differences (i.e. 0·15) and the total number of items (i.e. 30) that the final stage of the test is concerned. If the two distributions are very similar, then at no point will a large difference appear, while such a large difference will be apparent if there is a reasonable degree of dissimilarity between the two sets of

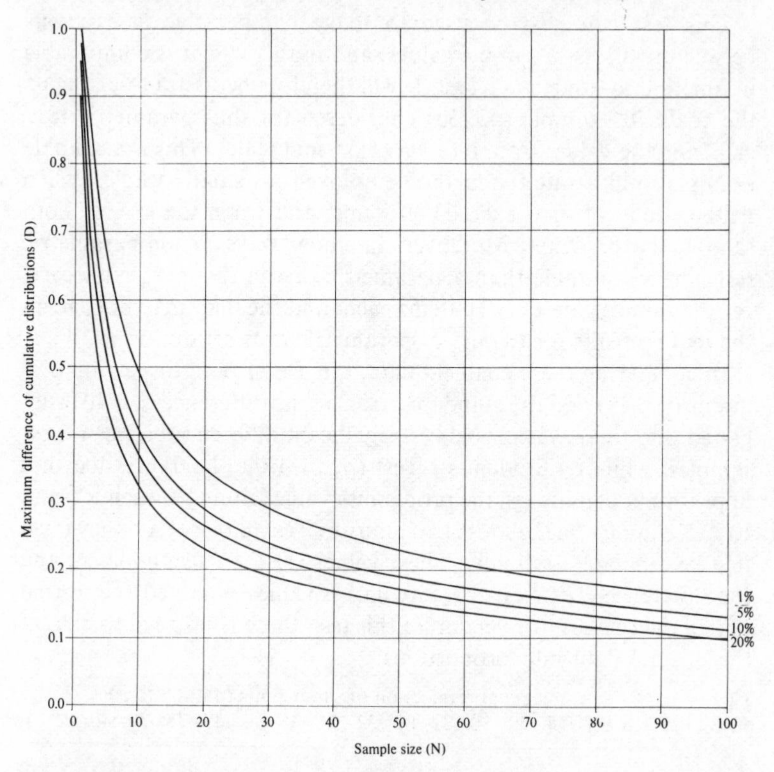

Figure 32. The Kolmogorov-Smirnov one sample test

values. The actual significance of any such difference, however, is also related to the number of items, i.e. the size of the sample, such that the smaller the sample the larger must be the greatest difference for any particular significance level to be reached. The statistical significance of the results of the analysis can be read from the graph provided in Fig. 32. It will be seen that for the present example, with a maximum difference of 0·15 for a sample size of 30, the null hypothesis must be accepted whichever of the given critical limits is used.

Thus, as it is highly probable that the null hypothesis is correct, one can assume that the particular month does not differ radically from the overall period as regards the frequency of daily rainfall amounts.

Two Sample Problems

This test can also be used to make comparative assessments between two sets of sample values, and in this way it is comparable in function to Student's *t* Test. It will therefore be useful to use again the coalfield example (p. 133) considered for that parametric test, changing the values from interval to ordinal scales. This is a suitable example to illustrate the method employed for small samples, when neither body of data exceeds 40 items, and when the size of both samples is the same. Moreover, in many ways a non-parametric test is more suitable than a parametric one in this particular case, for the samples are only 10 items each, and the data may not satisfy the requirements for the use of parametric tests set out on p. 132.

In comparing these coalfield data, the use of both dispersion diagrams (p. 135) and the standard error of the difference (p. 140) suggested that there was probably a significant difference between these samples, while by Student's *t* Test (p. 143) the difference does not appear to be significant, the probability value falling between 5% and 10%. To apply the Kolmogorov-Smirnov Test to these data they must first be tabulated ordinally. These values are then accumulated, and the differences between the cumulative values obtained. Note that unlike the one sample version of this test, there is no need to change the ordinal data into proportions.

Coal-	Annual coal production of sample pits (million tons)								
field	·25–·26	·27–·28	·29–·30	·31–·32	·33–·34	·35–·36	·37–·38	·39–·40	·41–·42
A	2	3	1	1	1	1	1	0	0
B	0	2	1	0	2	2	1	1	1
Accumulate									
A	2	5	6	7	8	9	10	10	10
B	0	2	3	3	5	7	8	9	10
Difference									
	2	3	3	4	3	2	2	1	0

As with the one sample test, it is only the greatest difference with which one is now concerned, and this (4, in this case) is referred to the appropriate probability tables (see Table XXIII) in relation to the size of the samples used (10 items in this example). In Table XXIII critical maximum differences at the 1% and 5% levels are given for samples of up to 40 items in size, but with different critical values depending on whether it is a *one-tailed* or a *two-tailed* test that is involved. The difference between these is simple, but needs understanding. In this example, the null hypothesis which has been tested can be put as:

H_0: there is *no* difference between the two sets of sample data for coalfields A and B.

If this is proved to be unlikely, the alternative hypothesis may be *either*

H_1: there *is* a significant difference between the two coalfields (but which is the greater is not considered), *or*

H_2: coalfield B has significantly greater values than coalfield A (i.e. the *direction* of difference is specifically involved).

In all the testing so far carried out it has been the first of these alternatives (hypothesis H_1) which has been concerned, and this requires a two-tailed test, in that one is interested equally in whether A is bigger than B and in whether B is bigger than A. If either is true then one is equally satisfied. In the present example, a maximum difference of at least 8 is required to establish a significant difference at the 1% level, i.e. to reject the null hypothesis with no more than a 1 % chance of error, and of 7 for one at the 5% level. As the actual maximum difference is only 4, the null hypothesis must be accepted, and therefore it is unlikely that a significant difference exists between these two sets of sample data. The conclusion thus agrees with that obtained by the Student's *t* Test rather than with the other two less stringent tests that were also applied to these data.

If the appropriate alternative to the null hypothesis is H_2, however, then a *one-tailed* test is required, for the possibility of coalfield A being greater than coalfield B is not a relevant consideration. As a result, a rather smaller maximum difference *in the appropriate direction* is required for any given probability level, but even in this case a difference of 6 is necessary for no more than a 5% chance of error when rejecting the null hypothesis, and this does not occur.

The method of operating this test with larger samples is basically the same, but critical probability limits are derived rather differently;

moreover, with samples of above 40 items there is no longer the need for both samples to be of the same size. If, in the above example, the sample for coalfield A had consisted of 50 items and that for coalfield B of 60 items, but the proportions of these falling into each production category had been the same as previously, one could tabulate and calculate as follows:

Coal-field	Annual coal production of sample pits (million tons) ·25–·26	·27–·28	·29–·30	·31–·32	·33–·34	·35–·36	·37–·38	·39–·40	·41–·42
A	10	15	5	5	5	5	5	0	0
B	0	12	6	0	12	12	6	6	6
Accumulate									
A	10	25	30	35	40	45	50	50	50
B	0	12	18	18	30	42	48	54	60
Proportions									
A	0·2	0·5	0·6	0·7	0·8	0·9	1·0	1·0	1·0
B	0·0	0·2	0·3	0·3	0·5	0·7	0·8	0·9	1·0
Difference									
	0·2	0·3	0·3	0·4	0·3	0·2	0·2	0·1	0·0

The only difference here compared to small samples is that after accumulation the values are converted to proportions, for with different sample sizes it would otherwise not be possible to obtain a valid 'difference' value. It is then necessary to compare the largest difference (0·4) with critical values related to the size of samples involved. Smirnov has indicated that, for a two-tailed test, the 5% level can be calculated by $1{\cdot}36\sqrt{\dfrac{n_1 + n_2}{n_1 . n_2}}$, the 1% level by $1{\cdot}63\sqrt{\dfrac{n_1 + n_2}{n_1 . n_2}}$ and the 0·1% level by $1{\cdot}95\sqrt{\dfrac{n_1 + n_2}{n_1 . n_2}}$. In the present example, where $n_1 = 50$ and $n_2 = 60$, the relevant values are:

$$5\% \text{ level} = 0{\cdot}2604$$
$$1\% \text{ level} = 0{\cdot}3121$$
$$0{\cdot}1\% \text{ level} = 0{\cdot}3734$$

Clearly, the null hypothesis of no difference between the two sets of data can be rejected with no more than a 0·1% chance of error, so that the difference that exists is a highly significant one statistically.

Table XXIII

Kolmogorov–Smirnov Two Sample Test (small samples)—critical maximum difference values for given sample sizes

Sample size (N)	One-tailed test 5% level	One-tailed test 1% level	Two-tailed test 5% level	Two-tailed test 1% level
3	3	—	—	—
4	4	—	4	—
5	4	5	5	5
6	5	6	5	6
7	5	6	6	6
8	5	6	6	7
9	6	7	6	7
10	6	7	7	8
11	6	8	7	8
12	6	8	7	8
13	7	8	7	9
14	7	8	8	9
15	7	9	8	9
16	7	9	8	10
17	8	9	8	10
18	8	10	9	10
19	8	10	9	10
20	8	10	9	11
21	8	10	9	11
22	9	11	9	11
23	9	11	10	11
24	9	11	10	12
25	9	11	10	12
26	9	11	10	12
27	9	12	10	12
28	10	12	11	13
29	10	12	11	13
30	10	12	11	13
35	11	13	12	
40	11	14	13	

(From S. Siegel, *Nonparametric Statistics for the Behavioral Sciences*, 1956, Table L.)

Thus, as has been seen everal times previously, if two exactly similar results are obtained, one with a small sample and one with a larger sample, the statistical significance of that result is far greater for the larger sample.

These non-parametric tests, and the many others that have not been considered here, mean that it is possible to test the statistical significance of quantitative conclusions, even when the data are in qualitative form. Which of the many tests should be used in any particular case, however, will depend on the nature of the data, the degree of accuracy required, the purpose for which the analysis is being made, and sometimes simply on the matter of preference. In the long run, it is only experience which will provide a sound guide when making such a choice between the various possible methods of testing the significance of the differences between sets of sample data.

CHAPTER 11

THE PROBLEM OF CORRELATION

In the previous three chapters methods have been presented by which the differences between sets of data can be tested as regards their statistical significance. The aim in all these cases was to assess whether the differences that were observed could have occurred by chance sufficiently frequently for some doubt to be cast on the validity of the apparent differences, or whether the probability of their having happened by chance was so slight that these observed differences could legitimately be accepted as justified and significant. In all these cases it was the overall characteristics of the various sets of data that were under consideration rather than the detailed characteristics and their changes. In many problems, however, there is the need to compare sets of data in terms of the extent to which a change in one is or is not reflected by a change in the other set. This necessarily implies that the individual items of the two sets of data coexist either in time or space, such that the possibility of interrelated changes can be considered. In such a problem an index is required that reflects the degree to which changes in direction (+ or −) and magnitude in one set of data are associated with comparable changes in the other set. An index of this sort is provided by what is termed the *Product Moment Correlation Coefficient*, and in the following pages this coefficient will be employed in the study of several problems.

Calculation of the Product Moment Correlation Coefficient

A simple case may be provided by comparing ten years of cereal yields for two districts of a country, as are set out below. It will be

	Years										
Districts	1	2	3	4	5	6	7	8	9	10	Average
x	25	26	34	25	24	28	27	29	28	29	27·5
y	21	17	35	21	19	22	26	22	26	26	23·5

Cereal yields in bushels per acre

seen that production varies from year to year in each district, that these variations are not always the same for the two districts, and that the average yields around which yearly values vary are themselves different. To try to assess from this by simple inspection the extent to which fluctuations in the yields of one district are reflected in those of the other district would lead to no more than a generalized impression, and in more complex situations even that would not be possible.

The fact that these fluctuations take place about different mean values increases the problem of comparison, but this can be eliminated by calculating the amount and direction by which each item differs from its respective average. The results are set out in Table XXIV under the headings $(x - \bar{x})$ and $(y - \bar{y})$. This tabulation does give a clearer picture. It can be seen, for example, that in seven of the ten years the two sets of data differ from their respective means in the same direction, though rarely by the same amount; in the other three years one district has above-average yields while the other has yields below the average.

Table XXIV

Tabulation of data for calculating the product moment correlation coefficient

x	y	$(x - \bar{x})$	$(y - \bar{y})$	$(x - \bar{x})(y - \bar{y})$ +	−
25	21	−2·5	− 2·5	6·25	
26	17	−1·5	− 6·5	9·75	
34	35	+6·5	+11·5	74·75	
25	21	−2·5	− 2·5	6·25	
24	19	−3·5	− 4·5	15·75	
28	22	+0·5	− 1·5		0·75
27	26	−0·5	+ 2·5		1·25
29	22	+1·5	− 1·5		2·25
28	26	+0·5	+ 2·5	1·25	
29	26	+1·5	+ 2·5	3·75	
$\bar{x} = 27·5$	$\bar{y} = 23·5$			117·75 − = +113·5	4·25

To find from these data a value which, for any one year, will express the combined variation from the mean, the simplest method is to multiply the two separate deviations together, i.e. $(x - \bar{x})(y - \bar{y})$,

This product gives positive values in some cases and negative in others, and these are entered separately in the tabulation (Table XXIV). Thus under the *positive* values of $(x - \bar{x})(y - \bar{y})$ fall all those years in which the deviation is in the *same* direction in the two districts, whether this be above or below the average, while the negative values of $(x - \bar{x})(y - \bar{y})$ are for those years in which the two deviations are in opposed directions. This is a basic reason for multiplying these deviations rather than summing them. If then the separate values under $(x - \bar{x})(y - \bar{y})$ are summed, and the total of the negative values subtracted from the total of the positive ones, then the *total* deviation is obtained. In the example being considered, this can be seen from Table XXIV to be +113·5. If this is now divided by the number of pairs of values being compared, then the *average* deviation is obtained.
Thus,

$$\frac{1}{n}\Sigma(x - \bar{x})(y - \bar{y}) = \frac{+113 \cdot 5}{10} = +11 \cdot 35$$

This average of the products of the deviations of the two sets of data from their respective means is based on actual changes, and is known as the *co-variance* of these sets of data. Thus, whereas when finding the variance of one set of data the mean is obtained of the sum of the deviations *squared*, in this case the mean is obtained of the sum of the *product* of *two* deviations. This measure of the relationship between conditions *as they occur* can then be compared to the overall deviations about the mean divorced from the time-scale itself. For any one set of data this is represented by the standard deviation. Here, with two sets of data, the same process is carried out as in the calculation of the co-variance, i.e. instead of converting the standard deviation to the variance for the *one* set of data by squaring it, the *two* standard deviations are multiplied together. The co-variance is expressed as a proportion of this value, thus giving the product moment correlation coefficient,

$$\text{i.e. the correlation coefficient } (r) = \frac{\frac{1}{n}\Sigma(x - \bar{x})(y - \bar{y})}{\sigma_x . \sigma_y}$$

The possible values of this coefficient lie between +1 and −1, the former indicating that the two sets of data vary in the same direction and by the same amount on all occasions, while the latter indicates that although the amount of variation is always the same the direction of that variation is always opposed. Thus if the means for the two sets of data were the same, as also were the standard deviations, i.e. $\bar{x} = \bar{y}$ and $\sigma_x = \sigma_y$, while a perfect correlation existed all the way, then the following transpositions in the formula could be made

$$r = \frac{\frac{1}{n}\Sigma\,(x - \bar{x})(y - \bar{y})}{\sigma_x . \sigma_y} = \frac{\frac{1}{n}\Sigma\,(x - \bar{x})(x - \bar{x})}{\sigma_x . \sigma_x} = \frac{\frac{1}{n}\Sigma\,(x - \bar{x})^2}{\sigma_x{}^2}$$

The top line of this is the expression for the variance, i.e. $\sigma_x{}^2$ so that

$$\frac{\frac{1}{n}\Sigma\,(x - \bar{x})^2}{\sigma_x{}^2} = \frac{\sigma_x{}^2}{\sigma_x{}^2} = +1$$

Equally, if a perfect inverse relationship existed a similar calculation would yield $r = -1$. In nearly all cases, however, actual values for r will lie within these limits.

Thus, in the example of crop yields begun above, the correlation coefficient would be obtained as follows:

$$r = \frac{\frac{1}{n}\Sigma\,(x - \bar{x})(y - \bar{y})}{\sigma_x . \sigma_y} = \frac{+11{\cdot}35}{2{\cdot}73 \times 4{\cdot}8} = \frac{+11{\cdot}35}{13{\cdot}1} = +0{\cdot}87$$

The co-variance value has already been calculated, while the reader can check the standard deviation values by any of the methods outlined in Chapter 3. This coefficient of +0·87 clearly implies that there is a high degree of positive correlation between these two districts in terms of fluctuations in cereal yields. Whenever values increase in one district there is a distinct tendency for them to increase also in the other, though this tendency is neither absolute nor of uniform magnitude. In general terms, coefficients of between +0·5 and +1 and between −0·5 and −1 are fairly significant, while if values lie between −0·5 and +0·5 then little significant correlation is to be expected. If a value of zero is obtained, this indicates that the two sets of data fluctuate completely independently of each other and that no correlation exists at all. To be safe in making any of these interpretations, however, the statistical significance of the correlation coefficient should always be tested by Student's t Test, to assess the probability of it having occurred by chance. This theme will be taken up at greater length on p. 197. Furthermore, great care must always be taken in interpreting correlation coefficients. This value of +0·87, for example, does *not* indicate *why* this relationship exists; it does not prove that the same causes have produced the same results, for there may well have been different factors at work producing these changes in the two areas. All it does is to indicate the degree of statistical relationship between the observed values—explanations must be sought by further work. Moreover, it should be apparent that, as this coefficient is largely determined in terms of variance and standard deviation values, it does assume a normal frequency distribution in the data; if this is not so, transformation to normality is an important prerequisite (see pp. 52–58).

Alternative Methods of Calculation

Nevertheless, some indication as to whether there is likely to be some valid relationship for which an explanation needs to be found is itself a valuable aid and guide. It helps to prevent explanations being put forward for relationships that are more apparent than real, and also indicates what is likely to be the most fruitful line of further research. As with all these statistical methods, it is a means to an end, not an end in itself. This being so, it is desirable to keep to a minimum the labour involved in calculation for this coefficient. This can be effected by a method very similar to that adopted for the standard deviation on pp. 27–31, for as has just been indicated the various components of the formula for the coefficient have much in common with the standard deviation and variance values.

It was shown on p. 28 that the formula for the standard deviation

$$\sigma = \sqrt{\frac{\Sigma (x - \bar{x})^2}{n}}$$

could be rewritten as

$$\sigma = \sqrt{\frac{\Sigma x^2}{n} - \bar{x}^2}$$

Equally the co-variance element in the calculation of the correlation coefficient can be altered from

$$\frac{1}{n} \Sigma (x - \bar{x})(y - \bar{y}) \quad \text{to} \quad \frac{\Sigma xy}{n} - \bar{x}.\bar{y}$$

This enables more rapid calculation, especially if the two standard deviations are also calculated by the shorter method, so that the formula becomes

$$r = \frac{\frac{\Sigma xy}{n} - \bar{x} \, . \, \bar{y}}{\left(\sqrt{\frac{\Sigma x^2}{n} - \bar{x}^2}\right)\left(\sqrt{\frac{\Sigma y^2}{n} - \bar{y}^2}\right)}$$

Such an equation can lead to large numbers being involved, and if considerable numbers of calculations are to be effected, it is essential that such mechanical computational aids as desk calculators (preferably electrically, if not electronically, operated) should be used.
In this case, with the problems involved in handling large numbers thus minimized, further manipulation of the correlation coefficient formula is possible, and it can be applied directly to the original data. The formula which perhaps requires the least number of individual calculations on a desk calculator is:

$$r = \frac{\Sigma xy - \frac{\Sigma x \, . \, \Sigma y}{n}}{\sqrt{\left(\Sigma x^2 - \frac{(\Sigma x)^2}{n}\right)\left(\Sigma y^2 - \frac{(\Sigma y)^2}{n}\right)}}$$

where x and y represent the original data.

Below are set out such data for annual (October to September) rainfall over, and run-off from, the River Etherow, for the period 1937–1938 to 1952–1953. A causal relationship between the amount of rainfall over an area and the amount of run-off from that same area is to be expected, and it is frequently of value to be able to express the degree of relationship in numerical terms. For this purpose the product moment correlation coefficient is of great value.

Table XXV

Annual rainfall over, and run-off from, the River Etherow (Oct. 1937–Sept. 1953)

Rainfall (in.) (x)	Run-off (in.) (y)
46·4	31·9
63·0	46·8
48·8	34·2
60·1	47·5
50·6	35·2
57·5	40·5
55·5	41·3
57·0	43·5
60·8	44·8
48·3	38·5
59·0	39·1
41·0	26·5
66·7	46·5
56·4	43·4
58·3	40·9
55·7	41·3

Using the last formula given, the following summations need be obtained:

	Σx	Σy	Σx^2	Σy^2	Σxy
rainfall/run-off values	885·1;	641·9;	49,634·03;	26,255·63;	36,044·77.

If these values are fed into the formula, it will be found that

$$r = \frac{536{\cdot}66}{\sqrt{671{\cdot}4 \times 503{\cdot}4}} = \frac{535{\cdot}66}{\sqrt{337{,}982{\cdot}76}} = \frac{535{\cdot}7}{581{\cdot}4} = +\,0{\cdot}92$$

As was to be expected, this value indicates a very high degree of positive correlation between annual rainfall and annual run-off for this drainage basin. However, this must not be interpreted to mean that 92% of the variations in run-off can be determined from the rainfall data. This latter characteristic is instead represented by the *coefficient of determination*, which is the correlation coefficient squared (r^2). Thus it is really 84·64% of run-off variations that can be so determined ($r^2 = 0{\cdot}92^2 = 0{\cdot}8464$), so that the proportion of these variations that cannot be assessed or determined from the rainfall data is not a mere 8% but almost twice that amount (almost 15·5%).

Yet a further value of studies such as this is that isopleth maps can be drawn based on correlation coefficients. Thus in Fig. 33, such

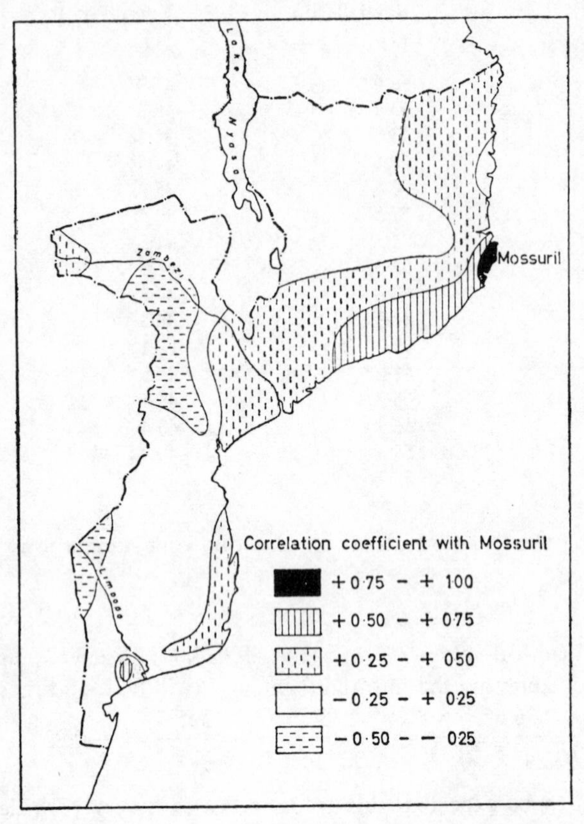

Figure 33. The correlation of annual rainfall in Moçambique with that at Mossuril

coefficients have been calculated for annual rainfall between Mossuril in Moçambique and all other climatological stations in that territory. From the resultant values isopleths are interpolated, thus providing a map showing the degree of correlation between annual rainfall at Mossuril and that of the rest of Moçambique.

Correlation Significance Test

The methods of calculating this coefficient have thus been illustrated at some length, and the kinds of problems to which the method can be applied have also been shown. In all such cases, however, there is always the possibility that the coefficient obtained could have occurred 'by chance', i.e. that its significance is suspect because of the probability of a chance occurrence. Therefore the correlation coefficient must be tested to see whether or not a chance occurrence of this magnitude is likely, as a result of the size of the sample or set of data analysed. This can be done by the use of the Student's t distribution, using the following formula:

$$t = \frac{r.\sqrt{n-2}}{\sqrt{1-r^2}}$$

where $n =$ the number of pairs of data studied, and where the degrees of freedom are $(n - 2)$. The null hypothesis that is postulated is that there is no correlation between the two variables, i.e. H_0 is $r = 0$.

In the first example in this chapter, where two sets of crop yields were compared, the necessary values were $r = +0{\cdot}87$ and $n = 10$. These can be introduced into the formula for Student's t, with the sign of the correlation coefficient always being taken as positive, simply for the sake of convenience. So it is seen that

$$t = \frac{0{\cdot}87 \times \sqrt{10-2}}{\sqrt{1-0{\cdot}87^2}} = \frac{0{\cdot}87 \times \sqrt{8}}{0{\cdot}493} = \frac{2{\cdot}46}{0{\cdot}493} = 5{\cdot}0$$

The degrees of freedom in this case are simply $n - 2 = 10 - 2 = 8$. By referring this t value and the degrees of freedom to the Student's t graph in Fig. 30, it can be seen that if the rejection begins at the 5%, 1% or even the 0·1% level, the null hypothesis can be rejected. In fact, it can be rejected with no more than a 0·1% chance of being wrong, and the inverse hypothesis that there *is* a correlation between the two sets of data can be accepted, i.e. this coefficient is highly significant statistically. This is even more true in the case of the rainfall run-off example, a statement which can readily be checked by the reader introducing the following value into the formula for Student's t:

example of run-off correlated with rainfall $r = +0{\cdot}92$ $n = 16$

To save the calculation of these significance levels in every case, a graph has been prepared (Fig. 34) from which they can be read directly. Thus it will be seen that if only 10 pairs of items are compared, giving but 8 degrees of freedom, then the correlation coefficient must be either above $+0{\cdot}69$ or below $-0{\cdot}69$ before it can be considered as statistically significant even at the 5% level. On the other hand, if about 60 pairs of items are compared, then a coefficient as low as + or $-0{\cdot}25$ is statistically significant at this level. If, however, a high degree of significance is required (i.e. at the 0·1% level) then the coefficient values must be markedly higher, e.g. with 40 degrees of freedom r must be greater than $+/-0{\cdot}5$, a value

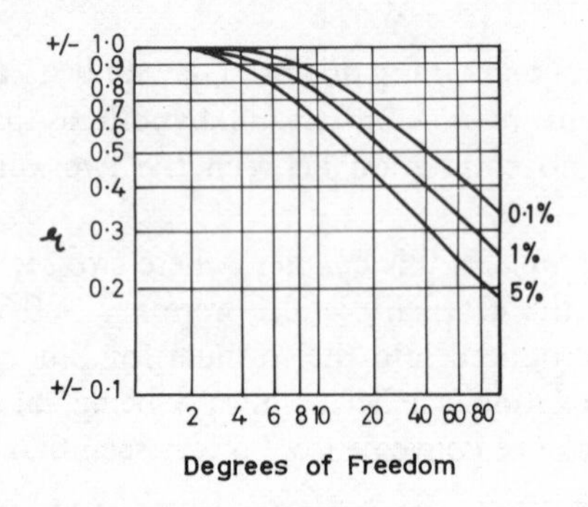

Figure 34. Graph of significance levels for correlation coefficients using Student's t distribution

which was suggested on p. 192 as being necessary as a general overall guide. Moreover, this means that in Fig. 33, when $n = 20$ and the degrees of freedom $= 18$, only those areas with a coefficient greater than $+0{\cdot}6$ have a significant correlation with Mossuril at the 1% level. Moreover, in terms of the coefficient of determination (r^2), this means that no more than 36% of the rainfall variations at such a station could be determined or assessed from those at Mossuril.

Spearman's Rank Correlation Coefficient

Even with the various means of shortening the calculations, this product moment correlation coefficient is not a value which is rapidly obtained. At times, therefore, it is convenient to use a rather different coefficient which is based not on actual values but rather on the relative *rank* of the values, i.e. where they occur in order of magnitude. Apart from this providing a quicker method of assessing correlation, there are many occasions when *only* such rankings are available and actual values are not known. A number of such methods are available, and the one that is perhaps easiest to compute—*Spearman's Rank Correlation Coefficient*(r_s)—will be outlined below.

The sort of problem which may be considered in this way is shown by the following example. It may be known that for five industrial areas their relative orders of importance for (*a*) engineering in general and (*b*) car manufacture in particular are as listed below.

Industrial areas

	(i)	(ii)	(iii)	(iv)	(v)
Engineering	1	2	3	4	5
Cars	3	2	1	5	4

As can be seen, these five areas do not fall in the same order (or rank) for these two activities. On the other hand, the three more important and the two less important areas are the same in each case. It may therefore be useful to assess the degree of correlation there is between engineering in general and the manufacture of cars in particular. The first stage is to tabulate the data in terms of 'rank'; then obtain the difference between the two sets of data in each case (d), square these differences (d^2) and sum these squares (Σd^2)—this is set out below.

Engineering (rank)	Cars (rank)	d	d^2
1	3	2	4
2	2	0	0
3	1	2	4
4	5	1	1
5	4	1	1
		$\Sigma d^2 =$	10

This value is then used in the following formula, in which n is the number of pairs of occurrences being considered:

$$r_s = 1 - \frac{6 \Sigma d^2}{n^3 - n}$$

In the above example this will give a value of

$$r_s = 1 - \frac{6 \times 10}{5^3 - 5} = 1 - \frac{60}{125 - 5} = 1 - \frac{60}{120} = 1 - 0{\cdot}5$$

$$r_s = +0{\cdot}5$$

This value suggests *some* relationship of a positive nature. If the significance of this value is checked from Fig. 34, however, the degrees of freedom being $n - 2 = 5 - 2 = 3$, then it will be appreciated that this value is *not* significant statistically at any of the given levels.

The limits of this coefficient are again +1 and −1. Thus if the degree of correlation were to be perfect and positive, with the ranking the same in each group, then the values for d in the above calculation would all be 0, as would therefore be both d^2 and Σd^2. As a result, the value $\frac{6 \Sigma d^2}{n^3 - n}$ would equal 0, so that r_s would equal $1 - 0 = +1$. If, on the other hand, the correlation were perfect but negative, the following would be the case.

Set *a*	Set *b*	*d*	d^2
1	5	4	16
2	4	2	4
3	3	0	0
4	2	2	4
5	1	4	16
		$\Sigma d^2 =$	40

$$r_s = 1 - \frac{6 \Sigma d^2}{n^3 - n} = 1 - \frac{6 \times 40}{5^3 - 5} = 1 - \frac{240}{120} = 1 - 2 = -1$$

Thus the formula is designed to ensure that +1 and −1 are the largest values that can be returned, so that in this way it is comparable to the product moment correlation coefficient.

There is also the question, however, as to whether or not it gives the same answer as does the more complicated method, or at least one which closely approximates to it. This can be tested by reworking the rainfall run-off data previously presented on p. 195. The values given on that page must first be put in ordinal form, i.e. in rank order, as shown below:

rainfall	run-off	d	d^2
15	15	0	0
2	2	0	0
13	14	1	1
4	1	3	9
12	13	1	1
7	10	3	9
11	7·5	3·5	12·25
8	5	3	9
3	4	1	1
14	12	2	4
5	11	6	36
16	16	0	0
1	3	2	4
9	6	3	9
6	9	3	9
10	7·5	2·5	6·25
			$\Sigma d^2 = 110{\cdot}5$

When two or more items occupy the same rank, the normal procedure is to allocate to each of them the average of the rank values that would have been assigned if no ties occurred. Thus the items seventh and sixteenth of the run-off column are joint seventh in rank; if ties had not occurred one of these would be seventh and one eighth, so here each of them is allocated a rank value of 7·5. This procedure is satisfactory so long as rank ties are infrequent; if they are common, however, further correction factors have to be introduced and modified formulae used. Details of this problem will not be considered here.

In this example this presents no real difficulty, however, and the sum of the squares value obtained above can be entered into the rank correlation coefficient formula:

$$1-\frac{6\Sigma d^2}{n^3-n}=1-\frac{6\times 110{\cdot}5}{16^3-16}=1-\frac{663}{4080}$$
$$=1-0{\cdot}16=+0{\cdot}84$$

This differs to some extent from the value obtained by the previous method, which gave $r = +0{\cdot}92$, but it is clearly an answer of the same order of magnitude, which provides an acceptable generalization at considerably less work.

Furthermore, on testing for significance from the graph in Fig. 34 it can be seen that this value lies beyond the 0·1% level, i.e. the null hypothesis can be safely rejected with less than 0·1% probability of being wrong, and the coefficient is statistically highly significant. With this ranking method, however, this testing of significance can only satisfactorily be carried out if n is not less than 10 (as is the case here). Moreover, as this coefficient is based only on rank and not on actual values, it is not a fully efficient index of correlation. The considerable shortening and simplification of the calculations involved, however, render it of great value especially for obtaining a generalized estimate of correlation, quite apart from the fact that in many cases only rank may be available for analysis.

The earlier warning concerning care in interpretation must be reiterated here at the end of this chapter on correlation. Such methods as those outlined are meant to be useful tools. They do *not* exempt the geographer from the necessity to think in a logical and sensible manner. It may well be quite possible to obtain a high correlation coefficient of statistical significance between two sets of conditions which clearly have nothing to do with each other—perhaps, for example, between coal production in Britain and the number of penguins in Antarctica in the same years! No one would try to suggest that a causal relationship exists between these two despite any correlation coefficient that may be obtained. In other cases, however, it may be more difficult to decide whether or not statistical correlation implies causal relationships.

Finally, it must be stressed that there are also a number of statistical assumptions that underlie the use of correlation techniques, and anyone using these methods for basic research purposes must bear these in mind. For discussion of these, however, reference should be made to more advanced books on correlation, for the concepts involved go far beyond the level of an introductory text such as this.

CHAPTER 12

REGRESSION LINES AND CONFIDENCE LIMITS

In many studies the calculation of a correlation coefficient, in any of the ways outlined in the previous chapter, may be sufficient in itself. This may indicate the most profitable lines for further research, or provide the data from which maps of iso-correlation may be drawn. In other cases, however, it may be desirable to take the analysis a stage further by calculating the value that might be expected for one set of data if some given value occurs in the other set. This could be done by separate calculations each time, but it is more effective to draw on a graph the line that represents the relationship between the two sets of data. The requisite values can then be read off as required. Such a graphical representation is equally valuable as a descriptive device to illustrate the form of the relationship between the two variables.

Straight-line Regression for Two Variables

In the case of a perfect positive correlation between two sets of data the individual values would be distributed as shown in Fig. 35*a*. They would all fall on a straight line, and this line could be drawn through the points without any further calculations. This, in effect,

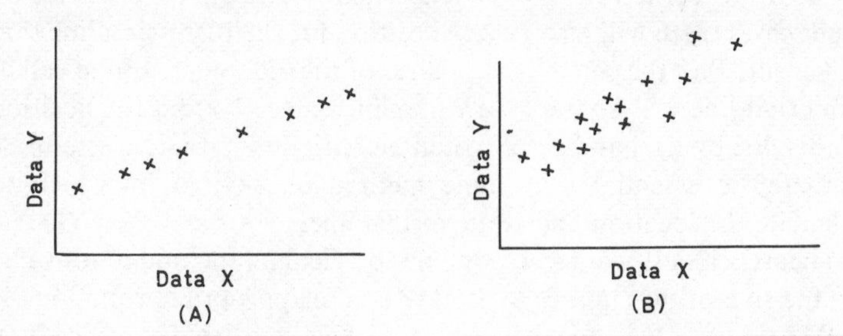

Figure 35. Graphs illustrating differing degrees of correlation and relationship

is a functional relationship which allows of no minor deviations from this straight line and which implies that any change of a given magnitude in one set of data must necessarily be associated with an exactly comparable change in the other set. Such a relationship is rarely found in the problems which confront geographers. Instead there is likely to be at best some sort of correlation, the degree of it being reflected by the coefficients outlined earlier. With correlation of this sort the distribution of the actual values on a graph will be comparable to that shown in Fig. 35*b*. There is clearly some sort of relationship, but it is neither regular nor clear-cut. The insertion of a line which summarizes this relationship, and from which can be assessed the most likely value of one variable when the other variable is known, is the best that can be done in this case, for there is no *one* value which *must* occur at any given point on the graph. Rather there are various possibilities, and what is required is a line that will give the closest approximation to the relationship at all stages.

Such a line as this is known as a 'regression line'. Unlike the situation when a functional relationship occurs, i.e. when $r = +1$ or -1 (Fig. 35*a*), it is not possible to insert a line by eye with any guarantee of accuracy, for such a visual insertion could be no more than guesswork. In obtaining the regression line by calculation, the idea is to ensure that the sum of the squares of the differences of the individual observed values from the line is at an absolute minimum. This is known as the method of 'least squares'. It may be visualized as being akin to ensuring that the *variance* of the individual values in relation to the regression line is the smallest value it can possibly be. Clearly, in the case of the functional relationship expressed by $r = +1$ (see Fig. 35*a*) there will be no deviations of actual values from the regression line for it passes through all the points. In all other cases there will also be *one* position for the regression line that will ensure that the sum of the squares of the differences of the values from that line will be the *lowest* possible value. To find the position of this line by trial and error would be both difficult and wasteful. It is therefore essential that some method be devised by which to calculate the location and slope of this line.

Theoretically it would be possible to calculate the minimum value for the sum of the squares by setting up the appropriate equation for each pair of values being considered. This is a lengthy procedure, however, and it is more convenient to apply a formula which gives the same result with much less labour. This formula requires not only the correlation coefficient but also the average and standard deviation values for the two sets of data. These have all been calculated for the correlation coefficient itself and if the first example in Chapter 11 be reconsidered for this purpose, the following values obtain:

$\bar{x} = 27{\cdot}5 \quad \bar{y} = 23{\cdot}5$
$\sigma_x = 2{\cdot}73 \quad \sigma_y = 4{\cdot}8$
$r = +0{\cdot}87$

The formula to be used is written as follows:

$$y - \bar{y} = r.\frac{\sigma_y}{\sigma_x}.(x - \bar{x})$$

in which y is the dependent variable and x the independent variable, i.e., the value of y is assumed to depend on the value given to x, the latter varying freely and not depending on y. In other words, dependent variable (y) differs from the average of its set of data($\bar{y}$) by the same amount as the known value x differs from *its* average ($\bar{x}$) modified by (i) the ratio of the two standard deviations, which express the overall spread of values about their respective averages and (ii) the correlation coefficient, which expresses the degree of the actual relationship unit by unit.

In the present example this becomes:

$$y - 23{\cdot}5 = 0{\cdot}87 \times \frac{4{\cdot}30}{2{\cdot}73} \times (x - 27{\cdot}5) = 1{\cdot}53\,(x - 27{\cdot}5)$$

$$y = 1{\cdot}53x - 42{\cdot}0 + 23{\cdot}5$$

Thus the regression of y (dependent) upon x (independent) is expressed by $y = 1{\cdot}53x - 18{\cdot}5$

In this equation, the value $+1{\cdot}53$ is known as the *regression coefficient* and expresses the degree of slope of the regression line (Fig. 36), in terms of the units of measurement for the two variables involved, i.e. it expresses the number of units of change in y per unit change of x. The value $-18{\cdot}5$ is the *base constant* and indicates the point of origin of the line, i.e. the value of the variable y when the variable x is zero.

By inserting values for x into this equation, the appropriate values for y can be obtained. Only two such values are required because the regression line is a straight line. Thus if $x = 27{\cdot}5$ then

$$y = 1{\cdot}53x - 18{\cdot}5 = (1{\cdot}53 \times 27{\cdot}5) - 18{\cdot}5 = 23{\cdot}5$$

These values for x and y (27·5 and 23·5 respectively)are the *average* values for the two sets of data. So in fact only *one* value really needs calculating since the other one is provided by the two average values. The second value in this case can be when $x = 30$, so that

$$y = 1{\cdot}53x - 18{\cdot}5 = (1{\cdot}53 \times 30) - 18{\cdot}5 = 27{\cdot}4$$

Even this calculation is not required if the zero value for the known variable x is included on the graph, for the base constant then provides the second value.

From these two values (the calculated or base constant ones and the averages) it is possible to draw the regression line of y on x, as has been done in Fig. 36. This line describes the general variation of y as x varies; it does not, however, describe the general variation of x as y varies. The formula is designed to ensure that the lowest value is obtained for the sum of the squares of the deviations of the y values from the line. The same line will only also yield the lowest

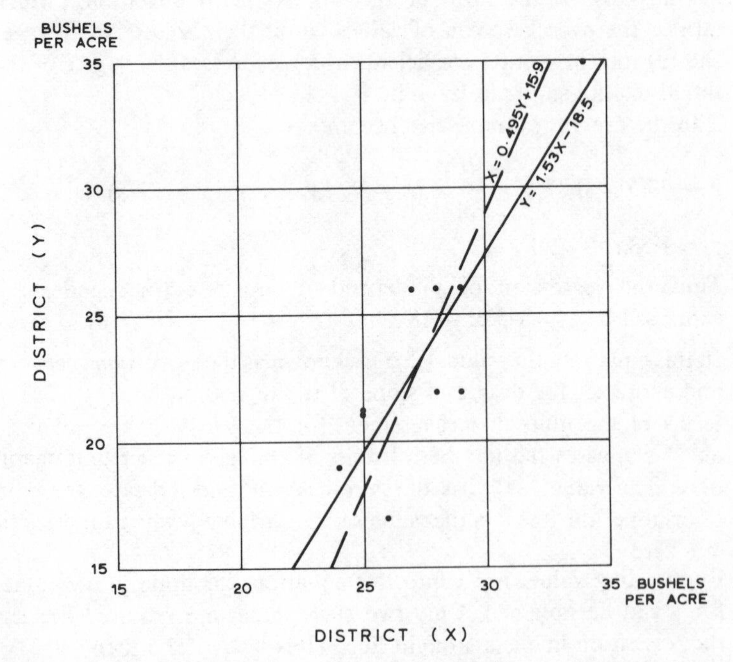

Figure 36. Regression lines for the relationship between cereal yields for two districts

sum of the squares for the x values if $r = +/-1$. In all other cases, therefore, it is necessary to calculate a *separate* regression line from which to assess x from y so that the lowest sum of the squares of the deviations of the x values from the line is obtained.

This is calculated by the same formula, such that

$$x - \bar{x} = r.\frac{\sigma_x}{\sigma_y}.(y - \bar{y})$$

In the present example this would give the following values:

$$x - 27{\cdot}5 = 0{\cdot}87 \times \frac{2{\cdot}73}{4{\cdot}80} \times (y - 23{\cdot}5)$$

$$x = 0{\cdot}495y + 15{\cdot}9$$

As in the previous case, the insertion of values for y will yield the appropriate values for x. Again, however, the two average values ($x = 27{\cdot}5$ and $y = 23{\cdot}5$) give one of the points and only *one* value of y need be inserted.

Thus, if $y = 20$ then $x = (0{\cdot}495 \times 20) + 15{\cdot}9 = 25{\cdot}8$

Equally, if $y = 0$ is included on the graph, then $x = +\ 15{\cdot}9$, i.e. the base constant.

This regression line, from which x can be assessed from y, has also been entered on Fig. 36. It can be seen that it differs from the one from which y values can be assessed. The angle of difference between these two regression lines reflects the relative size of the correlation coefficient. When it is $+/-1$ then the two lines coincide; when it is 0 then the two lines are at right-angles to each other; all other values of r give lines which differ from each other between these extreme limits.

In the present example the correlation was positive, and as a result the regression lines rise from left to right. If the correlation were to be negative then the lines would rise from right to left instead. For example, if the following data of crop yields and altitude are analysed, it will be found that $r = -0{\cdot}95$ and $y = 32 - 0{\cdot}01x$. The appropriate regression line is shown in Fig. 37. In both these examples, however, it must be stressed that these regression lines are only *best estimates* of the relationship between the two variables; equally the value for the dependent variable which this gives is only a best estimate. No more than this can be obtained, for with an

Altitude in feet (x)	Yield in bushels per acre (y)
100	30
200	30
500	31
700	24
800	26
1,000	23
1,400	13
1,500	17
1,800	14
2,000	12

imperfect relationship there cannot be *one* answer which *must* be right.

A further cautionary comment must be made concerning the regression coefficient and base constant. The actual magnitude of both of these reflects not only the relationships involved in the data, but also the *units of measurements* used. If these change, so too do these coefficients—a point to remember as metrication occurs!

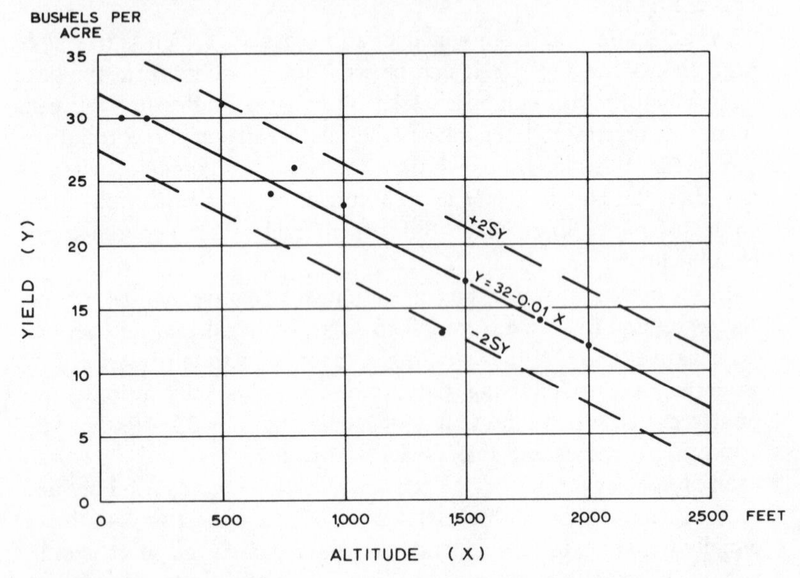

Figure 37. Regression line and confidence limits of yield on altitude

Standard Errors and Confidence Limits

As indicated above, it is desirable to be able to calculate the standard error of regression estimates, so that the range within which actual conditions are likely to fall can be assessed with some accuracy. This standard error of the estimate of the dependent value (e.g. of y) is expressed by the term (Sy) and is calculated by the formula

$$Sy = \sigma_y . \sqrt{1 - r^2}$$

It should be noted that the size of the standard error of the estimate is related to two things—it is directly related to how variable are the data for the dependent variable, and inversely related to the size of the correlation coefficient.

With this value obtained, the arguments presented several times earlier are again applied, i.e. that there is a 95% probability that actual values will differ from the regression line value by *not more than* twice the standard error, and that the probability of values differing by more than this amount is only 5%. This means that by the appropriate calculations *confidence limits* can be obtained in relation to the estimated values indicated by the regression analysis.

In terms of the first example in this chapter (p. 205) the standard error of the estimate can be found as follows. If it is the value of x that is being estimated, then the standard error $Sx = \sigma_x . \sqrt{1 - r^2}$

i.e. $Sx = 2{\cdot}73\sqrt{1 - 0{\cdot}87^2} = 2{\cdot}73\sqrt{1 - 0{\cdot}76} = 2{\cdot}73\sqrt{0{\cdot}24}$
$= 2{\cdot}73 \times 0{\cdot}49 = 1{\cdot}34$
Then $2\,Sx = 2{\cdot}68$

This means that when $y = 20$ then $x = 25{\cdot}8$ (see p. 207) $+/-2{\cdot}68$, with a 95% probability. In other words, there is a 95% probability that the value of x will lie between 23·1 and 28·5. Although such an answer to the query 'what will yields be in "District x" when they are 20 bushels per acre in "District y"?' may not seem as precise as saying bluntly 25·8 bushels per acre, it *is* more accurate and justified. Furthermore it reflects the somewhat variable relationship which is clearly apparent in Fig. 36. For this example it is equally possible to calculate the standard error of the estimate for values of y, when it is values of x that are known. In this case the calculations are as follows:

$$‘S_y = \sigma_y . \sqrt{1 - r^2} = 4{\cdot}80\sqrt{1 - 0{\cdot}87^2} = 4{\cdot}80 \times 0{\cdot}49 = 2{\cdot}35$$

Then $2\ Sy = 4{\cdot}70$

Thus when $x = 30$, then $y = 27{\cdot}4$ (see p. 206) $+/-4{\cdot}7$, i.e. y will lie between 22·7 and 32·1 bushels per acre, with a 95% probability.

Quite apart from calculating such a standard error for any given assessment it is possible to construct lines on the same graph as the regression line which will enable the ‘confidence limits’ to be read off at a glance. The 95% confidence limits have been entered on Fig. 37 which shows the regression line for crop yields on altitude, from the second example in this chapter. The regression line itself was expressed (p. 207) as

$$y = 32 - 0{\cdot}01x$$

while the standard error of this estimate of y becomes

$$Sy = \sigma_y . \sqrt{1 - r^2} = 7{\cdot}1\sqrt{1 - (-0{\cdot}95)^2} = 7{\cdot}1 \times 0{\cdot}312 = 2{\cdot}22$$

$$2\ Sy = 4{\cdot}44$$

Therefore along each line of altitude points were placed, 4·44 bushels above and below the regression line, and these values (one set above the regression line and one set below it) were joined together to give the 95% confidence limits. In this way it can be seen that at an altitude of 500 ft. there is a 95% probability that cereal yields will be between 23·56 and 32·44 bushels per acre. These are wide limits but reliable ones. More restricted limits indicating the range of values occurring with a 68% probability could be obtained by placing the limits only one standard error from the regression line. On the other hand, yet wider limits of 99·7% probability could be obtained if *three* standard errors were to be used.

Further Considerations

This whole series of calculations, by which a regression line and confidence limits are calculated for two variables between which a certain degree of correlation exists, will now be repeated for the third example used in Chapter 11, i.e. for the relationship between rainfall and run-off. In this way the repetition of the methods will help to reinforce the outline presented earlier. Moreover, it will also allow a number of alternative calculation procedures to be introduced.

The data from which the regression line may be calculated are

$\bar{x} = 55{\cdot}32 \quad \sigma_x = 6{\cdot}48 \quad \bar{y} = 40{\cdot}12 \quad \sigma_y = 5{\cdot}61 \quad r = +0{\cdot}92$

These values must then be inserted into the formula for the regression of y (run-off) on x (rainfall) to obtain the required regression coefficient and base constant, i.e.

$$y - \bar{y} = r.\frac{\sigma_y}{\sigma_x}.(x - \bar{x})$$

$$y - 40{\cdot}12 = 0{\cdot}92 \times \frac{5{\cdot}61}{6{\cdot}48} \times (x - 55{\cdot}32)$$

$$y = 0{\cdot}80(x - 55{\cdot}32) + 40{\cdot}12 = 0{\cdot}8x - 44{\cdot}26 + 40{\cdot}12$$

$$\underline{\underline{y = 0{\cdot}8x - 4{\cdot}14}}$$

For the two points required for the drawing of a regression line, one is provided by the two averages, i.e. when $x = 55{\cdot}32$ then $y = 40{\cdot}12$. The other is obtained by substitution in the expression for y

i.e. if $x = 60$, then $y = (0{\cdot}8 \times 60) - 4{\cdot}14$
$= 48 - 4{\cdot}14 = 43{\cdot}86$

These two points have been plotted on Fig. 38 and the regression line drawn along with the plotted values for each of the sixteen pairs of observations.

In this and all the other cases considered, it has been assumed that there has been some need for the prior computation of the correlation coefficient. This may not always be true, however, and it may simply be the regression equation that is required, as an expression of a generalized relationship between the dependent and the independent variable. In such cases it is not necessary to calculate the correlation coefficient, for it is possible to proceed directly to the regression equation from the same basic tabulations as are needed for the correlation coefficient. Thus on p. 195 are presented the initial summations from which to obtain the sums and averages of the rainfall and run-off data, the sums of the squares of these values,

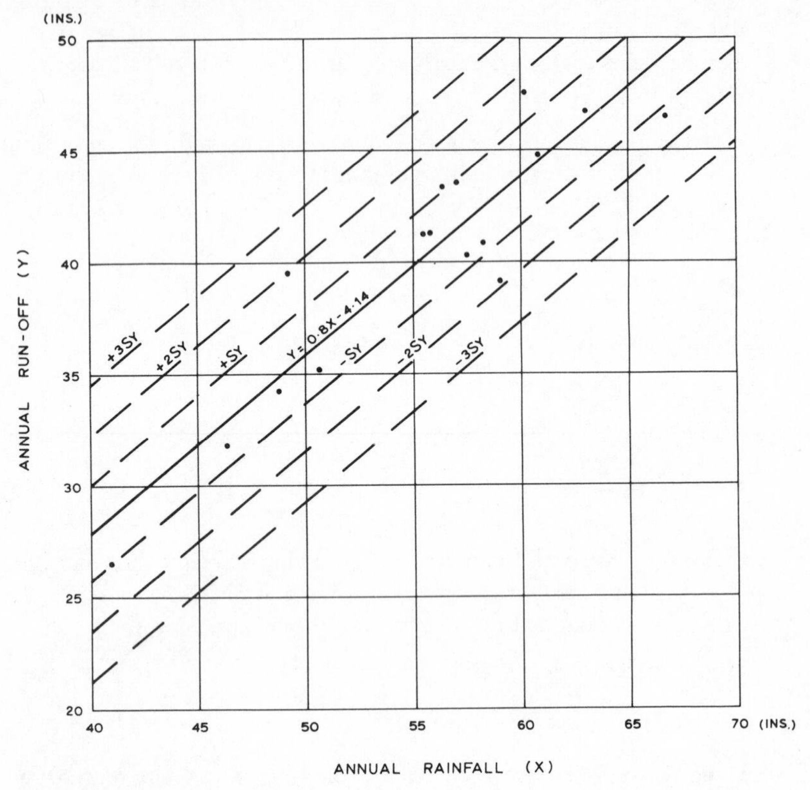

Figure 38. Regression line and confidence limits for the assessment of annual run-off from annual rainfall for the River Etherow

and the sums of the cross-products between the two. These initial summations were:

	Σx (independent)	Σy (dependent)	Σx^2	Σy^2	Σxy
rainfall/run-off values:	885·1	641·9	49,634·03	26,255·63	36,044·77

To obtain the regression equation, two values must be calculated, namely the regression coefficient and the base constant (p. 205).

$$\text{regression coefficient} = \frac{\text{sum of the cross-products}}{\text{sum of the squares of the independent variable}}$$

$$= \frac{\Sigma xy - \dfrac{\Sigma x . \Sigma y}{n}}{\Sigma x^2 - \dfrac{(\Sigma x)^2}{n}}$$

These values have also been previously calculated (p. 195) so that:

$$\text{regression coefficient} = \frac{535{\cdot}66}{671{\cdot}4} = \underline{\underline{+0{\cdot}8}}$$

From this the base constant is obtained:

base constant = average of the dependent variable—(average of the independent variable times the regression coefficient)

$$= \bar{y} - \text{regression coefficient}. \bar{x}$$

$$= 40{\cdot}12 - (0{\cdot}8 \times 55{\cdot}32) = 40{\cdot}12 - 44{\cdot}26 = \underline{\underline{-4{\cdot}14}}$$

Thus the regression equation is seen to be the same as by the previous method, i.e. $y = 0{\cdot}8x - 4{\cdot}14$

When obtaining the correlation coefficient, however, it has proved possible to test, by means of Student's t, the statistical significance of the relationships observed. It is therefore essential that, when obtaining a regression equation without incorporating the correlation coefficient, this equation itself should equally be tested for statistical significance. Student's t test can again be used for this purpose, the t value being calculated by:

$$t = \frac{\text{regression coefficient} \sqrt{\text{sum of the squares of the independent variable}}}{\text{standard error of the estimate of the dependent variable}}$$

From the values for the present example, partly given above but also below for the standard error, this becomes:

$$t = \frac{0{\cdot}8\sqrt{671{\cdot}4}}{2{\cdot}2} = \frac{0{\cdot}8 \times 25{\cdot}9}{2{\cdot}2} = \frac{20{\cdot}72}{2{\cdot}2} = 9{\cdot}42$$

degrees of freedom $= n - 2 = 16 - 2 = 14$

On referring these results to Fig. 30, it will be seen that this regression equation is of an extremely high degree of statistical significance, just as the correlation coefficient had been previously. The line thus being calculated and tested, it is now safe to present it in graphical form.

If such a graph were to be used for assessing the probable run-off for given annual rainfalls, it would be desirable to indicate the range within which actual conditions are likely to occur with a given probability. In other words, it is desirable to enter on the graph the 'confidence limits' for such an assessment. These are obtained in the following way, if the correlation coefficient is known (p. 209).

Standard error of the estimate of $y = Sy = \sigma_y . \sqrt{1 - r^2}$

i.e. $Sy = 5{\cdot}61\sqrt{1 - 0{\cdot}92^2} = 5{\cdot}61\sqrt{1 - 0{\cdot}8464} = 5{\cdot}61\sqrt{0{\cdot}1536}$
$= 5{\cdot}61 \times 0{\cdot}39$
$= \underline{\underline{2{\cdot}2}}$

Also $2\ Sy = \underline{\underline{4{\cdot}4}}$ and $3\ Sy = \underline{\underline{6{\cdot}6}}$

If, on the other hand, the regression equation has been obtained without the correlation coefficient having been calculated, then a different formula can be used. The purpose of the standard error of the estimate is to express the extent to which the actual observations deviate from the generalization of the regression line. It can therefore equally be calculated by first working out the values for the dependent variable which the regression equation would yield, and then taking the difference between this estimate and the actual values. These differences are known as *residuals* (z), and the standard deviation of these residuals, expressing their deviation around the regression line, is another equivalent way of expressing the standard error of the estimate. Thus:
Standard error of the estimate = standard deviation of the residuals

$$= \sqrt{\frac{\text{sum of the squares of the residuals}}{\text{no. of items}}}$$

If the present example of rainfall/run-off relationships be considered, as in the tabular presentation below, this becomes a general formula of the type:

$$Sy = sz = \sqrt{\frac{\Sigma z^2}{n}}$$

Run-off values (p. 195) y	Run-off estimates ($y = 0{\cdot}8x - 4{\cdot}14$) y^1	Residuals $\lvert y-y^1 \rvert$ z	Residuals squared $(y-y^1)^2$ z^2
31·9	33·0	1·1	1·21
46·8	46·3	0·5	0·25
34·2	34·9	0·7	0·49
47·5	43·9	3·6	12·96
35·2	36·3	1·1	1·21
40·5	41·9	1·4	1·96
41·3	40·3	1·0	1·00
43·5	41·5	2·0	4·00
44·8	44·5	0·3	0·09
38·5	34·5	4·0	16·00
39·1	43·1	4·0	16·00
26·5	28·7	2·2	4·84
46·5	49·2	2·7	7·29
43·4	41·0	2·4	5·76
40·9	42·5	1·6	2·56
41·3	40·4	0·9	0·81
			$\Sigma z^2 = 76{\cdot}43$

$$Sy = \sqrt{\frac{\Sigma z^2}{n}} = \sqrt{\frac{76{\cdot}43}{16}} = \sqrt{4{\cdot}7769} = 2{\cdot}2$$

Clearly this is a much more lengthy procedure than using the formula incorporating the correlation coefficient. On the other hand, the calculation of residuals is often a valuable guide to further lines of research, and needs to be effected anyway.

This value for the standard error, by whichever method it is obtained, can then be used in relation to the two points from which the regression line is drawn, the following being the various limits of the confidence lines.

	68% prob.	95% prob.	99·7% prob.
If $x = 45$ then $y =$	29·7 – 34·1	27·5 – 36·3	25·3 – 38·5
and if $x = 55{\cdot}3$ then $y =$	37·9 – 42·3	35·7 – 44·5	33·5 – 46·7

These several values have also been entered on Fig. 38, thus giving three sets of confidence limits for this assessment of run-off from rainfall data. In this way a guide is given not only to the probable run-off from the catchment area but also to the likelihood with which such values will occur. These can be read off from Fig. 38, but it must be remembered that such values will only hold true if the *straight-line* relationship postulated applies for *all* rainfall ranges. There is here the possibility that as rainfall reaches very high values, e.g. about 80 in., then run-off values may deviate from such an hypothetical relationship. This is always a problem with regression lines, and it is only safe to apply them to the ranges of values on which the calculations are based. In this case the regression line and confidence limits should be satisfactory for falls at least between 40" and 65" and almost certainly between 35" and 70", i.e. within the likely range of values. Exceptional falls, whether they be high or low, may not be so adequately interpreted.

Before leaving this theme of regression equations related to two sets of variables, one further point must be briefly raised. It has been argued in previous pages that, because the data being analysed are sample data only, it is essential to test both the correlation coefficient and the regression equation by the t test to assess their statistical significance; also because any relationship is unlikely to be a perfect one, the standard error of the estimate must be calculated. Additional to these, however, it follows that the regression coefficient, which expresses the degree of slope of the regression line, is also only an estimate of the true regression line based on the total population, so that another separate sample of that same population is likely to yield a somewhat different estimate of the true regression coefficient. It is therefore desirable to obtain the standard error of any regression coefficient that is calculated from the available sample, as this will allow both an assessment of its statistical significance to be made and also an estimate of the range within which the true regression coefficient is likely to lie.

This is obtained as follows:

$$\text{S.E. regression coefficient} = \frac{\text{S.E. of the estimate of the dependent variable}}{\text{standard deviation of the independent variable}.\sqrt{\text{no. of items}}}$$

In the rainfall/run-off example which has been used above (see pp. 211 and 214 for values), this becomes:

$$\text{S.E. regression coefficient} = \frac{Sy}{\sigma_x . \sqrt{n}}$$

$$= \frac{2{\cdot}2}{6{\cdot}48\sqrt{16}} = \frac{2{\cdot}2}{25{\cdot}92}$$

$$= 0{\cdot}085$$

Thus, the regression coefficient of +0·8 has a standard error of 0·085, so that the true coefficient lies between +0·715 and +0·885 with a 68% probability, and between +0·63 and +0·97 with a 95% probability. Moreover, as the regression coefficient is 9·4 times as large as its standard error, it is obviously a highly significant one, and would be even if the t distribution were used to define the critical limits, as the sample size is less than 30 (compare p. 94). Similarly, if the regression equation for areal yields in Region x upon those in Region y (p. 207) be considered, i.e. $x = 0{\cdot}495y + 15{\cdot}9$, the standard error of the regression coefficient + 0·495 will be found to be 0·088. This coefficient is thus also of a high degree of statistical significance, being 5·6 times larger than its standard error.

Straight-line Regression for One Variable

The calculation of a regression line between two variables which have some correlation with one another is thus a fairly simple operation, the formulae presented and used above effecting a 'least squares' fit of the regression line to the data with a minimum of labour. In many problems for which a regression line would be useful, however, two correlated variables are not involved. Rather the data consist of only *one* variable, the occurrences of which are available for some regular interval either in space or time. The problem here is to construct a regression line that will express the relationship between changing location (in space or time) and changing magnitude of the occurrence. Thus if in the earlier example of crop yields varying with altitude the crop data had been obtained every 100 ft. instead of at irregular intervals, then a regression line of yields with altitude could have been obtained without first calculating the correlation coefficient. Again, in Chapters 2 and 3 several of the examples were based on either iron-ore production of four countries over a period of twenty years, or annual rainfall values at Bidston for a thirty-year period. In both cases these represent data for *one* variable, the values showing conditions at regular intervals. Also in both cases a semi-regular change of values with time could be expected as a distinct possibility, and such a change (if it does really exist) can be represented by a regression line. This theme here impinges on that of trends and fluctuations which will be considered at greater length in Chapter 13. Therefore, although the methods of calculating such a regression line will be outlined here, the implications of such a line in terms of trends will be left for consideration in the next chapter.

In calculating a regression line for data of this sort, two assumptions must be made. The first is that any relationship that exists holds true over the whole period or distance, while the second is that the relationship can be represented by one specific type of curve or line. It is therefore necessary to postulate, for example, that the most likely relationship is that which is represented by a straight line; in other cases, the logarithmic curve (p. 226) may be assumed to give the best fit to observed conditions. Whichever curve is assumed will control not only the calculations but also the conclusions that are likely to be drawn from the resulting graph. This fact must always be borne in mind.

If the rainfall data for Bidston, tabulated in Chapter 2 (p. 11), are used for the first example of this method, then it would seem reasonable to assume that if there *is* any change of values with time it may well approximate to a straight-line curve. This is not to argue that any such change will necessarily take place at a uniform rate throughout the period, but simply that as an *idealized curve* it is likely to be reasonably close to reality. In calculating such a linear relationship the aim is to assess the number of units by which the dependent variable (in this case, rainfall) changes for each unit change of the time or distance factor (in this case, successive years). As there is no steady functional relationship between time and rainfall, such a study can only provide an assessment, and again the regression line is drawn so as to ensure that the sum of the squares of the differences of the actual rainfall values from this line is at minimum, i.e. it is based on the 'least squares' method again.

If the years involved are listed uner (x) and the appropriate rainfall under (y), then the number of units (b) which y will increase per unit increase of x will be obtained by the formula:

$$b = \frac{\Sigma (x - \bar{x})(y - \bar{y})}{\Sigma (x - \bar{x})^2}$$

Table XXVI Calculation of a regression line for annual ıainfall at Bidston for the period 1901–30

Years x	Rainfall y	x^2	xy
1	25	1	25
2	26	4	52
3	34	9	102
4	25	16	100
5	24	25	120
6	28	36	168
7	27	49	189
8	29	64	232
9	28	81	252
10	29	100	290
11	25	121	275
12	30	144	360
13	26	169	338
14	26	196	364
15	27	225	405
16	25	256	400
17	31	289	527
18	32	324	576
19	29	361	551
20	33	400	660
21	22	441	462
22	26	484	572
23	31	529	713
24	33	576	792
25	28	625	700
26	29	676	754
27	35	729	945
28	29	784	812
29	25	841	725
30	36	900	1,080

$\Sigma x = 465 \quad \Sigma y = 853 \quad \Sigma x^2 = 9{,}455 \quad \Sigma xy = 13{,}541$

This formula will ensure that the resulting repression line will fit the 'least squares' requirement. If this formula is considered a little more carefully it will be seen that it is the same as that given earlier (p. 217) for calculating the regression coefficient between two variables, without first obtaining the correlation coefficient. What the formula implies is that if the difference of rainfall values from the rainfall average was unit for unit the same as the difference of the occurrence number from the average of the occurrence numbers, then $(x - \bar{x})$ would be the same as $(y - \bar{y})$. In such a case the expression $(x - \bar{x})(y - \bar{y})$ would be the same as $(x - \bar{x})^2$ so that the value of (b) in the above formula would be unity. This means that the amount by which the value of (b) differs from unity is controlled by the values of $(y - \bar{y})$. If these are larger than $(x - \bar{x})$ then (b) will be more than unity, while if they are smaller than $(x - \bar{x})$ then (b) will be less than unity. In this way it can be seen that the value of (b) is based on the relationship of the sum of the squares of $(x - \bar{x})$ and the sum of the *products* of $(x - \bar{x})$ and $(y - \bar{y})$. Thus the resulting regression line is located so that the sum of the squares of $(y - \bar{y})$ is kept to the minimum.

The calculation of the necessary values can be done directly from the raw data as outlined earlier (p. 213), in which case the formula for the regression coefficient becomes:

$$b = \frac{\Sigma xy - \dfrac{\Sigma x . \Sigma y}{n}}{\Sigma x^2 - \dfrac{(\Sigma x)^2}{n}}$$

The resulting tabulated summations are set out in Table XXVI, and can be fed into the equation as follows:

$$b = \frac{13{,}541 - \dfrac{465 \times 853}{30}}{9{,}455 - \dfrac{465 \times 465}{30}} = \frac{318{\cdot}5}{2{,}247{\cdot}5} = \underline{\underline{+0{\cdot}14}}$$

This means that for every unit change of (x), i.e. for each year's change, there will be a $+0{\cdot}14$ unit change of (y), i.e. a change of $+0{\cdot}14$ inches.

Having thus calculated the regression coefficient it is then necessary to obtain the base constant to complete the regression equation, Again the method outlined on p. 213 can be used, so that:

$$\begin{aligned} a = \bar{y} - b.\bar{x} &= 28{\cdot}43 - (0{\cdot}14 \times 15{\cdot}50) \\ &= 28{\cdot}43 - 2{\cdot}17 = 26{\cdot}26 \end{aligned}$$

The regression equation, expressing the relationship between annual rainfall and time at Bidston for the period 1901–1931, is therefore
$y = 0{\cdot}14x + 26{\cdot}26$
when x is the year after 1900 and y the rainfall in inches.

Two sets of values are needed to insert the regression line in Fig. 39

of which one set consist of the two averages, i.e. when $x = 15{\cdot}5$ then $y = 28{\cdot}43$. The other set can be obtained by inserting a value for x in the regression equation, and if this is 30 then $y = 30{\cdot}46$.

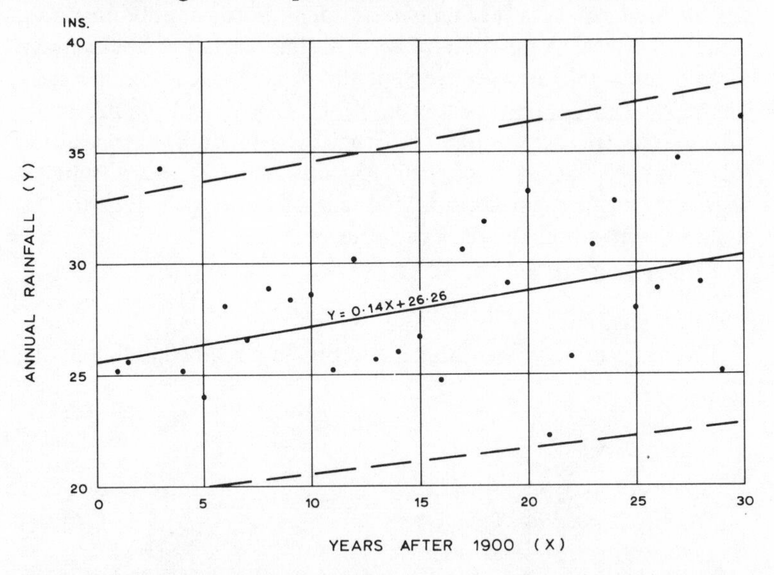

Figure 39. Regression line and confidence limits of annual rainfall at Bidston for the period 1901–1930

This line in Fig. 39 thus suggests that throughout the period under review (1901–1930), a slight overall increase in rainfall has occurred, though actual values differ quite markedly from the idealized values of the regression line. This theme of the differences between actual and idealized values will be considered further in Chapter 13.

Straight-line Regression for Spatial Change

This method of calculating a straight-line regression can also be applied to data which represent changes in space, with the observations taken at regular intervals. Suppose, for example, that remnants of a former cliff-line have been plotted over a considerable north–south extent of a westward-facing coastline. There are reasons to suppose that this area has been warped to a slight extent. However, the available data are in a variety of situations, some being at former headlands, others in bays or along estuaries, while the rocks on which they exist are themselves of varied character. As a result the heights of the cliff bases do not clearly indicate whether warping has taken place or not, apparently fluctuating indiscriminately. If the cliff-foot heights are available every mile in a straight north–south direction it would, however, be possible to calculate the regression line between these heights and distance, so that the possible trend can be seen. The postulated data are set out in Table XXVII, and the necessary calculations are there presented. From these it can be seen that the regression coefficient $b = -0{\cdot}1105$, i.e. that for every unit of distance (1 mile) southwards the cliff-foot heights decrease by 0·1105 ft. From this the regression formula is

$$y = 26{\cdot}11 - 0{\cdot}1105x$$

and this regression line is shown in Fig. 40.

It would also have been possible to obtain a regression line by the

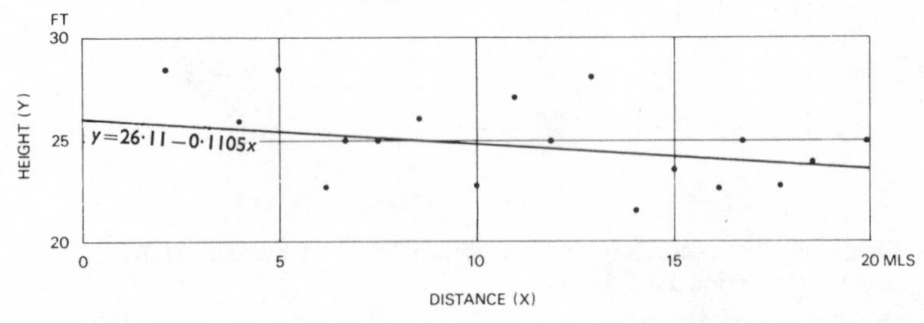

Figure 40. Regression line of cliff-foot heights on distance from north to south

methods outlined at the beginning of this chapter. Thus a correlation coefficient could have been calculated between height and distance, and the regression of height on distance obtained. This would necessarily have involved more calculation, though several advantages would have accrued from the results of this extra work. The correlation coefficient would have been $r = -0{\cdot}365$ and this could then be tested for significance. With a sample of only 20 values this does not reach the 5% level of significance, but if the sample were to be increased to 30 and the same degree of correlation held true, then this 5% level of significance would apply. The regression formula would be the same as in Fig. 40, i.e. $y = 26{\cdot}11 - 0{\cdot}1105x$, and various confidence limits could also have been calculated. This is not always done with the present method, often only a rough guide being provided by inserting limits at 25% of the regression line value above

Table XXVII

Calculation of a regression line for cliff-foot heights upon north-south distance

Distance in 1 mile units from N–S	Height above m.s.l. of cliff-foot		
x	y	x^2	xy
1	25	1	25
2	28	4	56
3	24	9	72
4	26	16	104
5	28	25	140
6	23	36	138
7	25	49	175
8	25	64	200
9	26	81	234
10	23	100	230
11	27	121	297
12	25	144	300
13	28	169	364
14	22	196	308
15	24	225	360
16	23	256	368
17	25	289	425
18	23	324	414
19	24	361	456
20	25	400	500

$\Sigma x = 210 \quad \Sigma y = 499 \quad \Sigma x^2 = 2{,}870 \quad \Sigma xy = 5{,}166$

Regression coefficient of y on x, i.e.

$$b = \frac{\Sigma xy - \dfrac{\Sigma x . \Sigma y}{n}}{\Sigma x^2 - \dfrac{(\Sigma x)^2}{n}} = \frac{5{,}166 - \dfrac{210 \times 499}{20}}{2{,}870 - \dfrac{(210)^2}{20}} = \frac{-73\cdot 5}{665\cdot 0} = \underline{\underline{-0\cdot 1105}}$$

Average values:

$$\bar{x} = \frac{210}{20} = 10\cdot 5$$

$$\bar{y} = \frac{499}{20} = 24\cdot 95$$

Base constant:

$$a = \bar{y} - b.\bar{x} = 24{\cdot}95 - (-0{\cdot}1105 \times 10{\cdot}5)$$
$$= 24{\cdot}95 + 1{\cdot}16 = \underline{\underline{26{\cdot}11}}$$

Regression equation:

$$y = 26{\cdot}11 - 0{\cdot}1105x$$

Points for regression line:

When $x = 10{\cdot}5$ then $y = 24{\cdot}95$
When $x = 20$ then $y = 23{\cdot}9$

and below the regression line itself (see Fig. 39 for an example of this). The accuracy of the values quoted here can be checked by the reader from the data and methods presented earlier, while the possibility of better confidence limits is discussed later on p. 247. The whole comparison stresses the fact that the type and value of the data that can be obtained by statistical analysis depends on the methods used and the amount of work put into the analysis. As a working rule the more complete the analysis, the more varied and reliable is the information that is obtained.

Curvilinear Relationships

In all these considerations of regression equations and lines, it has been assumed that the form of the relationship between the two variables is essentially that of a straight line. In other words, we have been concerned with the function that is shown diagrammatically in Fig. 35*a*, and which is expressed by a formula of the type:

$$Y = a \pm bX$$

where Y is the dependent variable, X is the independent variable, and a and b are constants (the former is the base constant and the latter the regression coefficient).

But in many problems of relationships between two variables, a straight line is *not* the best expression to be used. Rather, the relationship that exists has a closer approximation to some other non-linear function, so that the 'best-fit' regression line is in fact a curve which can be defined by a mathematical expression or equation different from that given above for a straight-line function. Probably the most commonly used of these in geographical work, at least for the less complex problems, are those based on either logarithmic functions or upon power functions, but a wide range of polynomial functions (quadratic, cubic, quartic, etc.) are also relevant in particular cases.

A number of such functions are illustrated in Fig. 41, from which a few generalizations can be made. Apart from the polynomial example, all the other functions *never* change their slope from positive to negative, or vice versa; the examples given here are all positive slopes, but in each case a negative slope could occur, with the Y values decreasing as the X values increase. In contrast, the polynomial functions, here represented by a typical parabola of the quadratic function, *do* involve at least one such change of direction. Another point to stress is that a relationship involving an *increasing rate of change* in Y per unit increase in X can be represented both by a power function (with the exponent or power value at least equal to unity—case 2 in Fig. 41) and by logarithmic functions when the *dependent* variable is expressed in logarithmic terms (cases 7 and 8 in Fig. 41). Equally a *decreasing rate of change* in Y per unit increase in X is presented by a power function (with the exponent less than unity, i.e. case 3 in Fig. 41) and by logarithmic functions when it is the *independent* variable that is in logarithms (cases 5 and 6 in Fig. 41).

These two types of curves are also possible when *both* the dependent and the independent variable are in logarithmic terms, for this is the same as the power function curves illustrated in Fig. 41. Thus the regression coefficient for the logarithm of the independent variable is the same as the exponent or power of that variable when logarithms are not used. For example:

$\log y = 2 \log x$ is the same as $y = x^2$; or $\log y = 0{\cdot}5 \log x$ is the same as $y = x^{0{\cdot}5}$

Moreover, logarithmic relationships may be expressed in terms of logarithms either to the base 10 or to the base e, the exponential function (see p. 80), i.e. in either normal or Naperian logarithms (or to any other base, if need be). Also, as these powers are in effect equivalent to regression coefficients for logarithmic relationships, they, too, owe part of their magnitude to the units of measurement used, so that changes in such units can lead to changes in the power function (see p. 208).

Thus, for any pair of related sets of data, it is possible to obtain a

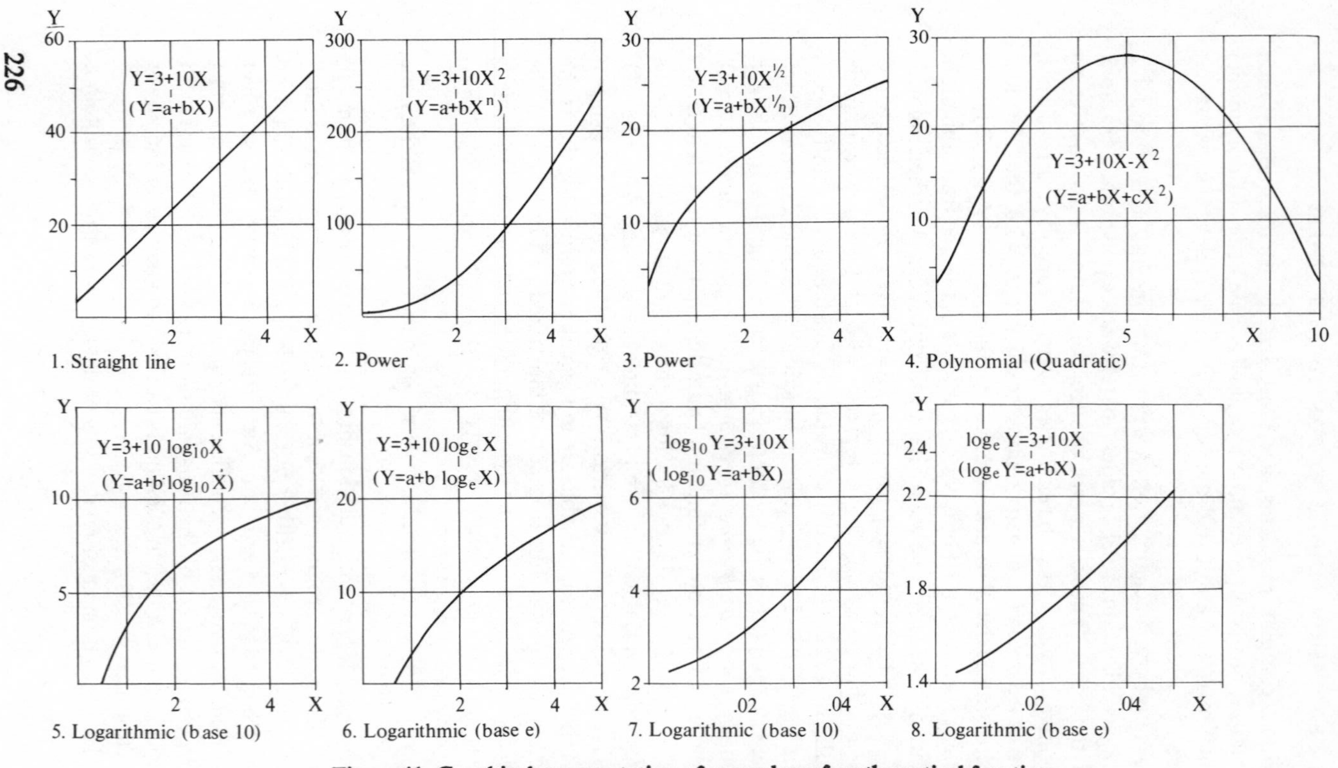

Figure 41. Graphical representation of a number of mathematical functions

large number of possible regression lines, each of which will provide the best fit to the original data in terms of the relationships assumed by the mathematical functions adopted. Which of these many possibilities really provides the best fit to the data, i.e. which function is the most suitable type of expression of the relationship that exists, can be assessed by calculating the residuals from the various regression lines and from this obtaining the standard error of the estimate (p. 215). The function which yields the smallest standard error is thus shown to provide the closest approximation to the real relationships. Thus the criticism that is at times raised against regression analyses that different results are obtained by assuming different functional relationships is only valid if these happened to incorporate exactly similar standard errors, which is highly unlikely.

The calculating of curvilinear regression lines can be an involved process, but as logarithmic functions are very commonly used in geographical studies, and as regression lines related to these raise few if any new calculation problems, the curvilinear regression problem can perhaps be most simply illustrated in these terms, with the unit of recording being at a regular interval.

In studies of population data, for example, it must always be remembered that the size of the population at one moment in time will affect its size at some later moment in time, just as it has itself been affected by the size of the population at some earlier period. As a result, population values do *not* always increase from one period to another by a uniform and constant *amount*. Instead, they often tend to increase by a uniform and constant *rate*. Thus the change with time is not *arithmetic* (as has been the assumption in earlier examples) but is rather *geometric*. In this way the increase is not expressed in such terms as 'a 10,000 increase per half-century' but rather as '*a twofold* increase per half-century'. Moreover, there may often be at least an element of this geometric increase in the case of industrial production values, while there is also frequently a geometric relationship between distance along the long-profile of a river and change of altitude.

To illustrate the characteristics of data which fit the logarithmic relationship the following simple example provides a suitable starting point. If four numbers, set out in succession, are

1; 3; 9; 27

it is clear that there is a threefold rate of increase from one value to the next, i.e. that there is a common *rate* of increase as distinct from a common *amount* of increase. With a more complex set of values this may not be appreciated so easily, especially if the rate of increase were not in terms of whole numbers. It is true that the relationship even in such a case could be arrived at by trial and error, but this is extremely slow and laborious with no guarantee of success. The difficulty partly arises for the very reason that the absolute difference between adjacent values is never constant—thus, in the present simple case these differences are 2; 6; 18. What happens, on the other hand, if the values involved are changed to logarithms? They then assume the values given below:

Original value	Logarithm (base 10)	Difference between successive logarithms
1	0·0	
		0·47712
3	0·47712	
		0·47712
9	0·95424	
		0·47712
27	1·43136	

Clearly, once the logarithms are considered instead of the original values, a constant *amount* of change is introduced again.

Once this has been done it is possible to calculate the necessary regression line by the same formula as before (pp. 217–224), using the logarithms of the values instead of the values themselves to obtain the 'logarithmic curve'. This is a perfectly legitimate device, but it does mean that care must be taken to interpret the results aright. As an illustration of the method the values given above, which are known to fit the logarithmic curve perfectly, can be examined, tabulating them as follows:

Items x	Values y	Log. of values $\log y$	x^2	$x.\log y$
1	1	0·0	1	0·00000
2	3	0·47712	4	0·95424
3	9	0·95424	9	2·86272
4	27	1·43136	16	5·72544

$\Sigma x = 10 \quad \Sigma \log y = 2{\cdot}86272 \quad \Sigma x^2 = 30 \qquad \Sigma x.\log y = 9{\cdot}54240$

Regression coefficient $= b$ (i.e. the log increase of y per unit increase of x)

$$b = \frac{\Sigma x.\log y - \dfrac{\Sigma x.\Sigma \log y}{n}}{\Sigma x^2 - \dfrac{(\Sigma x)^2}{n}} = \frac{9{\cdot}54240 - \dfrac{10 \times 2{\cdot}86272}{4}}{30 - \dfrac{10^2}{4}} = \underline{\underline{+0{\cdot}47712}}$$

Thus the same answer is obtained as in the simple tabulation, so it can be appreciated that this method yields the correct answers. A full application is probably best done in connection with a specific example in which the values only approximate to, and do not perfectly fit, the logarithmic curve.

Suppose that a study were being made of the colonization of an area of tidal flats by some particular plant species. A given section of that tidal area may be studied over a period of years, and the number of plants of the specific type occurring there is counted each year. In the first year the species has only just begun to colonize the area, and only three plants were to be seen. With natural regeneration, however, the numbers increase steadily, the values counted for the first six years of the study being as given in the table below. Clearly this increase is not linear, and considering the type of phenomenon being studied it may be expected that the logarithmic curve would provide a regression line which would fit the data more closely. The calculations, using the logarithms of the values for the number of plants, are set out in Table XXVIII.

Table XXVIII

Calculation of the logarithmic curve for the increase of plants over an area of tidal flats

Years x	No. of plants y	$\log y$	x^2	$x.\log y$
1	3	0·47712	1	0·47712
2	8	0·90309	4	1·80618
3	25	1·39794	9	4·19382
4	80	1·90309	16	7·61236
5	250	2·39794	25	11·98970
6	700	2·84510	36	17·07060
$x = 21$		$\log y = 9{\cdot}92428$	$x^2 = 91$	$x.\log y = 43{\cdot}14978$

$\Sigma x = 21 \quad \Sigma \log y = 9{\cdot}92428 \quad \Sigma x^2 = 91 \quad \Sigma x.\log y = 43{\cdot}14978$

Regression coefficient:

$$b = \frac{43{\cdot}14978 - \dfrac{21 \times 9{\cdot}92428}{6}}{91 - \dfrac{21^2}{6}} = \underline{\underline{+0{\cdot}48}}$$

Average values:

$$\bar{x} = \frac{21}{6} = \underline{\underline{3{\cdot}5}}$$

$$\overline{\log y} = \frac{9{\cdot}92428}{6} = \underline{\underline{1{\cdot}654}}$$

Base constant:

$$a = \overline{\log y} - b.\bar{x} = 1{\cdot}654 - (0{\cdot}48 \times 3{\cdot}5)$$
$$= -\,0{\cdot}026$$

Regression equation:

$$\underline{\underline{\log y = 0{\cdot}48x - 0{\cdot}026}}$$

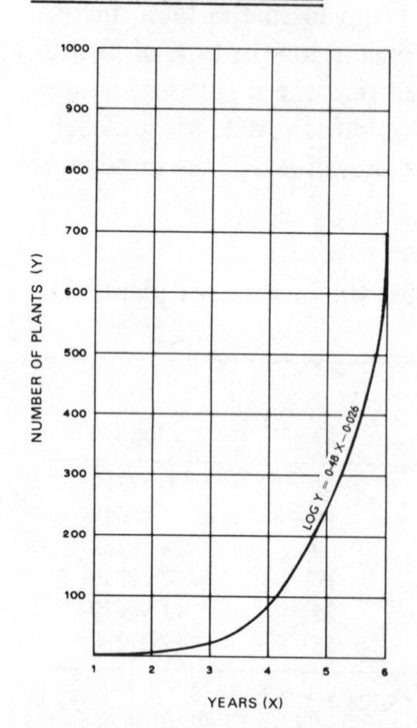

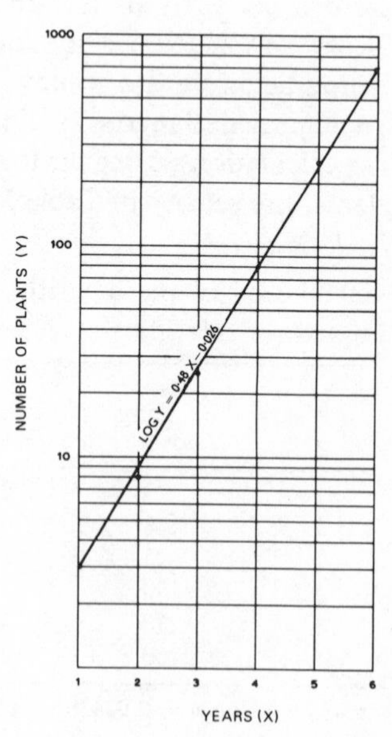

Figure 42. Regression line (logarithmic curve) on ordinary and on semi-logarithmic graph paper

As for the points from which to draw the regression line, the number of such points that are required depends on whether ordinary graph paper or semi-logarithmic graph paper is being used. In the first case, all of the values of x from 1 to 6 must be substituted in this formula in turn, so that each point is calculated. This will yield a curved line as is shown in Fig. 42*a*. If semi-logarithmic graph paper is being used, however, only two points are needed, of which one is provided by the two average values already calculated. This is because the construction of the graph paper ensures that a line showing a constant rate of increase will be plotted and drawn as a straight line (Fig. 42*b*). These two values could thus be

(i) when $x = 3{\cdot}5$ then $\log y = 1{\cdot}654$ (and $y = 45{\cdot}08$)
(ii) when $x = 6$ then $\log y = 2{\cdot}854$ (and $y = 714{\cdot}50$)

Logarithmic regression lines can, of course, also be calculated for the two-variable situation considered earlier in this chapter, either with or without the prior computation of a correlation coefficient. In such studies, it is necessary to convert the appropriate variable (or perhaps both variables) into logarithms before proceeding with the calculations outlined on pp. 203–217, and from this will result a regression line comparable to those in Fig. 41: again, semi-logarithmic paper will convert this to a straight line if only one of the variables was in logarithmic form. As for power functions, these can also be converted to straight-line graphs, this time by the use of double-logarithmic graph paper, i.e. with a logarithmic scale in both directions.

Thus many types of regression lines have been considered in this chapter. Which is to be used in any particular case must be decided by prior consideration of the data, the one chosen being that which most closely fits the data. The two forms used here were straight and logarithmic regression lines—others of greater complexity should be studied from more advanced texts if that is so desired. In all these cases, however, the specific purpose of the regression line is to express the relationship between data and location (or data and data) as precisely as possible, always bearing in mind the fact that there is not a perfect relationship between them. The regression line provides the closest fit, based upon the 'least squares' approach. Once prepared, it represents the relationship that exists in terms of the *available observations*, and it thus provides an illustration of relationships as they have existed or do exist. Prognostication *may* be carried out on the basis of such lines, but there is no necessary *statistical* reason why they should apply outside the data on which they are based. If the nature of the phenomenon under study renders this likely, however, e.g. in terms of rainfall and run-off, then these regression lines acquire a yet greater value and significance. Such considerations, which are related to trends and fluctuations, especially over time, are considered somewhat more fully in the following chapter.

CHAPTER 13

FLUCTUATIONS AND TRENDS

In all the problems which have so far been considered in this book, the aim has been to reduce or eliminate the detailed differences between one particular value and another, so that the overall characteristics can more readily be appreciated. Even in the case of correlation, where individual values were more directly considered, the purpose was to obtain *one* index which would summarize the full set of individual relationships. With some problems, however, the geographer must necessarily concern himself with the details of the changes from one individual value to another. This is so when the data consist of values which change in relation to changes in the time-scale. Thus it is possible that changes in production, in climatic conditions or in population values bear some relationship to such time-scale changes. This has already been partially indicated in the previous chapter, but even there the purpose of the regression lines was to present the *overall* change rather than the details of the actual changes.

The Simple Graph

When considering such details of change with time, i.e. when the fluctuations of a given set of values are being analysed, it is necessary to have recourse to graphical representation. If such fluctuations were found to occur with a clearly definable regularity then it would be possible to represent this by some mathematical expression. If, however, the fluctuations are of an irregular nature then such a mathematical summary can only be made at the expense of detail, and graphical illustration can give a clearer picture of conditions.

The simplest method of showing fluctuations is by means of a graph in which values of the phenomenon concerned are plotted against time and then these points joined by a continuous line. Such graphs are shown in Fig. 43 and Fig. 44. The former is for the annual rainfall data for Bidston, various characteristics of which have been assessed previously, while the second is for the output of crude petroleum by the U.S.A. for the twenty years 1937–1956, the values for which are set out in Table XXIX.

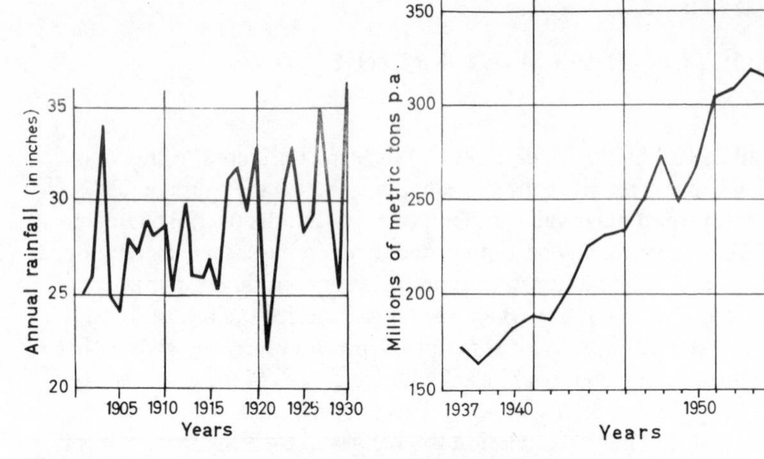

Figure 43. Fluctuation in annual rainfall at Bidston, 1901–1930

Figure 44. Fluctuation in U.S.A. crude petroleum production, 1937–1956

The pattern of change with time of petroleum production is readily apparent, the curve being almost universally upwards save on four occasions. Each of these falls lasted for only one year, and the picture as a whole is both uncomplicated and readily appreciated. On the graph of rainfall, however, this simplicity no longer applies. Values increase and decrease with apparent irregularity, and the definition of periods of rising or falling values, or of spells of wetter or drier years, becomes increasingly subjective. Moreover, if an attempt were

Table XXIX

U.S.A. crude petroleum production, 1937–1956, in millions of metric tons

Year	Production	Year	Production
1937	173	1947	251
1938	164	1948	273
1939	171	1949	249
1940	183	1950	267
1941	189	1951	304
1942	187	1952	309
1943	203	1953	319
1944	227	1954	313
1945	232	1955	336
1946	234	1956	354

to be made to compare such a graph of fluctuations with a similar graph for some other station, it would prove exceedingly difficult to pass any worthwhile judgment on whether or not the details of the fluctuations bore any relationship one to the other.

Running Means

With difficult cases such as this, there are other devices that may be used to simplify the task of judgment and assessment. The first of these aims at smoothing out the sharp and marked irregularities that can be seen in Figs. 43 and 44 so that only the major fluctuations are stressed and so need be considered. This can be effected by the calculation of 'running means'. This implies that if 'five-year running means' are being used, for example, then the first value will be the average of years 1–5; the second value will be the average of years 2–6; the third value will be the average of years 3–7, etc., until the final five years of the period. For the Bidston data the first two values would be as follows:

Years	Rainfall	First five-year mean	Second five-year mean
1901	25·19		
1902	25·57		
1903	34·42	26·87	
1904	25·18		27·45
1905	24·01		
1906	28·08		

Any number of years may be the basis for such a smoothing technique, but it must be borne in mind that *if* there were to be a regular periodicity in the fluctuations of *the same length as* the running-mean period, then such a regular fluctuation would not appear in the resultant graphs. It is therefore usually desirable to prepare such graphs for two periods of different lengths. These Bidston data have therefore been changed into both 'five-year' and 'ten-year' running means, and the respective graphs are shown in Fig. 45. In both, an overall though interrupted increase in rainfall values is indicated. On the basis of the five-year periods, values seem markedly to increase after the period 1913–1917 (mid-year 1915), although smaller fluctuations are seen to occur both before and after this period. From the values for the ten-year periods it would seem that rainfall increased after the decade 1913–1922, or possibly after 1904–1913, to a maximum in 1918–1927, while again smaller fluctuations are also apparent.

Such differences as these are, however, really differences between sample means. Therefore before any further reasoning or conclusions are based on these apparent differences, they should be tested by the methods outlined in Chapter 8 to assess whether they could well have occurred by chance, or whether they are statistically significant. One

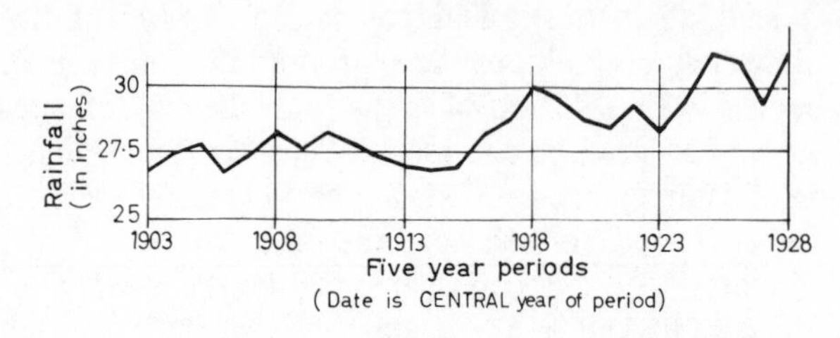

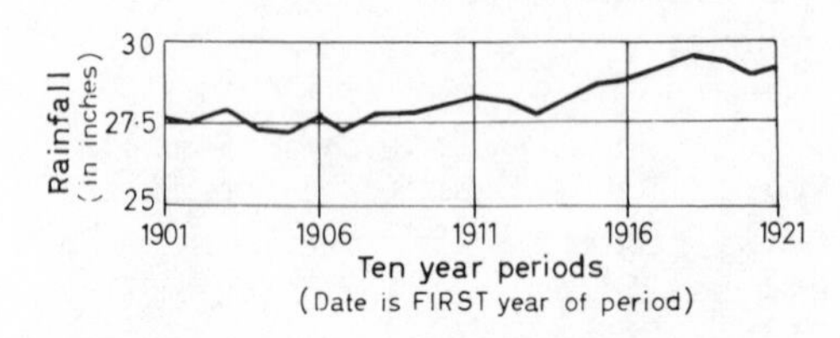

Figure 45. Graphs of running means for annual rainfall at Bidston, 1901–1930

possibility is to use the 'standard error of the difference' test, but as the size of the samples is relatively small it is better to apply Student's *t* Test. Thus in the case of the ten-year running means it would be desirable to test whether the difference between a dry decade (1904–1913) and a wet decade (1918–1927) is statistically significant or not.

The basic values of average and standard deviation required for the application of Student's *t* Test can be calculated from the data in Table I (p. 11). They are as follows:

	Decade	Sample average	Sample standard deviation
(*a*)	1904–1913	27·1	1·92
(*b*)	1918–1927	29·8	3·66

From these, Student's t can be calculated thus:

$$t = \frac{|\bar{a} - \bar{b}|}{\sqrt{\dfrac{s_a{}^2}{n-1} + \dfrac{s_b{}^2}{n-1}}} = \frac{27{\cdot}1 - 29{\cdot}8}{\sqrt{\dfrac{3{\cdot}68}{9} + \dfrac{13{\cdot}34}{9}}}$$

$$= \frac{2{\cdot}7}{\sqrt{0{\cdot}41 + 1{\cdot}48}} = \frac{2{\cdot}7}{\sqrt{1{\cdot}89}} = \frac{2{\cdot}7}{1{\cdot}375} = \underline{\underline{1{\cdot}96}}$$

The degrees of freedom are

$$(n_1 + n_2 - 2) = 10 + 10 - 2 = 18$$

and by reference to Fig. 30 it can be seen that these values do not quite reach the 5% level. Thus there is a probability of just more than 5% that a difference as great as this could have occurred by chance, so that it is not fully justified to argue that this difference is a probably significant one. However, if conditions at neighbouring stations indicated a change over the same period that was statistically significant, then it would be reasonable to treat a case such as this as falling in the same category—though still maintaining a certain element of possible doubt.

The application of such a test is not merely a nuisance imposed by statistical requirements. It can rather be a positive help in focusing attention on those differences which really are statistically significant and in avoiding the tendency to explain smaller differences which are quite likely to be solely chance occurrences. Thus if the five-year running means were to be considered, the difference apparent in Fig. 45 between 1913–1917 (average value 26·82″) and 1923–1927 (average value 31·11″) would at first sight appear to be an important one. By applying Student's t Test, and having calculated that the respective best estimates of the standard deviations from the two samples was 2·22″ and 2·80″, it can be found that $t = 1{\cdot}684$ with 8 degrees of freedom. From Fig. 30 this is shown to represent a difference between sample means that could have occurred by chance with a probability of greater than 10%. Thus in this case the apparent fluctuation involving a change in five-year means of the order of 4·37″ cannot be accepted as statistically valid, and further evidence must be sought before such a fluctuation should be accepted as a reasonable possibility.

Cumulative Deviations from the Mean

One difficulty with using running means is that even if a statistically significant change were to be established, it would not be possible to indicate exactly when such a change became effective. This renders comparisons rather difficult, while any attempt at assessing causal relationships from such graphs is equally hindered. Such difficulties are largely overcome if a different sort of graph is used instead. This graph is designed to show *cumulative deviations from the mean*, either in absolute or percentage terms. Only simple calculations are required for this. First the difference between each occurrence and the mean value is obtained, and these values are tabulated. The points on the graph are then calculated by progressively summing these differences, i.e. the first point is the difference between the first value and the mean; the second point is the sum of this difference and the difference between the second value and the mean; and so on to the end of the record. This is perhaps more clearly seen from Table XXX using the petroleum data for the U.S.A. given in Table XXIX.

Table XXX

Calculation of values for graphs of cumulative (percentual) deviations from the mean

Values	Difference from mean	Cumulative difference	% difference
(x)	$(x - \bar{x})$ (when $\bar{x} = 247{\cdot}4$)	$\Sigma (x - \bar{x})$	$\dfrac{\Sigma (x - \bar{x}).100\%}{\bar{x}}$
173	−74·4	− 74·4	− 30·05
164	−83·4	−157·8	− 63·7
171	−76·4	−234·2	− 94·8
183	−64·4	−298·6	−120·8
189	−58·4	−357·0	−144·4
etc.	etc.	etc.	etc.

Curves based on such calculations are presented in Fig. 46 and Fig. 47 for these petroleum data and for the Bidston rainfall data. In the case of the former, values are expressed as percentages of the mean, while in the latter they are shown as absolute values in inches. From the petroleum graph (Fig. 46) it can be seen that a series of lower-than-average years were followed, from 1947 onwards, by a series of above-average years. In Fig. 47, the dominance of drier-than-average years prior and up to 1916 is clearly seen, while the greater frequency of occurrence of wetter-than-average years after this date is also clear. It must be stressed, however, that actual position on the graph is irrelevant when an interpretation is being made in terms of rate and direction of change. The significant features are the *direction and angle* of slope of the graph. Whenever this rises it indicates an increase in values (even if this occurs where the graph reads −200%), while the steeper it rises the more rapid and marked that increase happens to be. Equally, however, if the rate at which the line falls gets less, then this indicates an increase in values

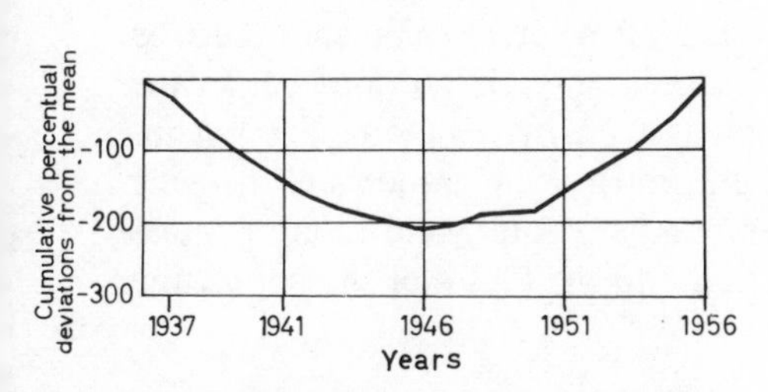

Figure 46. Graph of cumulative percentual deviations from the mean for U.S.A. crude petroleum production, 1937–1956

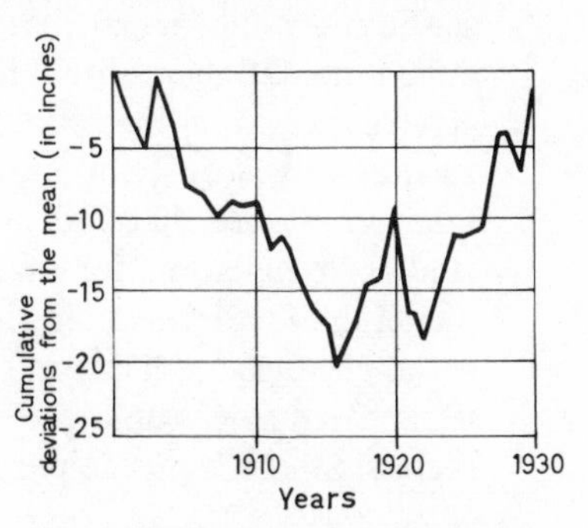

Figure 47. Graph of cumulative deviations from the mean for Bidston annual rainfall, 1901–1930

even though such increased values are still below the mean itself. Clearly the date at which a series of below-average conditions are replaced by a series of above-average conditions can be readily appreciated. On the other hand, a certain amount of practice is required for the ready interpretation of the graph in Fig. 46. This indicates a virtually continuous rise in values by the steadily decreasing rate at which the line falls and then its conversion to a line rising at an increasing rate.

Whichever of these methods is used the reason for using it is to represent the changes that have taken place with time. This may be desired simply to specify conditions at that one place or for that one commodity. At other times the purpose may include a comparison with the changes that have occurred elsewhere or in some other product. In neither case, however, can or should these methods be used to project beyond the actual period of the data. They are indicators of the past, not harbingers of the future. If such assessment is considered desirable, then the study of these detailed changes is best replaced by the regression lines which were outlined in Chapter 12. These regression lines are, in effect, trend lines which generalize the overall changes that have taken place. Even in these cases, however, great care should be taken to ensure that the factors that have caused this trend are likely to continue in the future, or—if the attempt is made to project back to the past—that they applied there too. Thus in the case of annual rainfall at Bidston, the facile assumption that the trend over 1901–1930 has always applied and will continue into the future would mean that in the twenty-first century the expected annual rainfall there would be over 40″, while at the beginning of the eighteenth century there would have been no rain at all! An absurdity such as this is only too apparent, but in other cases care must be taken to ensure that similar false reasoning is not applied. In the study of population, for example (see pp. 241–244 and Fig. 48), innumerable factors including health, nutrition, migration and changing social customs are likely to confound any forecast of future populations based solely on a projection into the future of the population regression line from the past.

Deviation from a Trend Line

The construction of regression lines to represent past trends can be of value in geography in another way. Being concerned with the variability of sets of data, the geographer is presented with a problem when the set of data itself includes a distinct trend throughout the period. In such cases, the calculation of variance and standard deviation values can be somewhat misleading, for they will be compounded of two elements, (i) the overall trend from the beginning to the end of the period and (ii) the variability of conditions from one occurrence to the next, which clearly occurs when the actual values do not perfectly fit the trend line. Thus in terms of the data on U.S.A. petroleum production, the overall trend reflects a steady increase throughout the period 1937–1956, but actual values nevertheless varied in relation to this trend. The calculation of variance values by the normal method for these 20 years may be useful as a device by which to summarize the characteristics of conditions over those particular 20 years. It should not be assumed, however, that it also fairly represents the longer series of data from which those 20 years were drawn. Such a variance is not the result of values varying at random about the mean value, but rather it is an abstraction which gives an inadequate picture of a set of data in which a consistent trend is occurring.

The same is true in terms of population values. Given a series of population data which consists of census returns at—say—ten-year intervals, it would be possible to calculate in the normal way the variance of these values in relation to the average of the body of data. However, because of the tendency for population to increase from one decade to another, it would also be possible to calculate the variance of the actual conditions in relation to those represented by an overall trend line. Such a value would reflect the combined influences of all those factors *other* than that of natural growth which the trend line (assuming that the logarithmic curve is used) would itself define. The variance or standard deviation, obtained in the normal way, would be dominated by the factor of natural growth, and these other factors—which may well be the important factors differentiating one area from another—would be largely obscured.

An example in terms of population data will help to clarify this approach, and illustrate the type of problem that is amenable to it. The following set of values could well represent the population of a small rural parish at ten-year intervals over a period of 70 years. By normal methods it can be calculated that the mean of these values is 512·5, that the best estimate of the standard deviation is 89 and that the coefficient of variation is 17·4%.

Decade (x)	Decadal returns (y)	
1	390	
2	435	
3	475	Suggested population
4	480	values for
5	500	a rural parish
6	550	
7	620	
8	650	

Clearly, however, there is a trend throughout this period which displays a continuous though variable increase, so that these deviation and variation values reflect not only fluctuations but also this tendency for continued growth. To separate these two elements it is desirable to construct a regression line that expresses this rate of growth and then to calculate the degree of fluctuation that occurs around this line. The form of the regression line will reflect the basic hypothesis concerning population growth. If it were to be the logarithmic curve, then the assumption would be that the major element of growth had been natural increase. If it were to be a straight line, then the assumed relationship would include some other basic factor (e.g. migration) acting concurrently with natural growth. These assumptions would necessarily affect the ultimate interpretation of any values of deviations from these trends that might be defined. In the present case either of these two hypotheses could be put forward, but for purposes of this example the dominance of natural increase will be assumed.

The first requirement is therefore to construct the appropriate regression line, and following the calculation procedures outlined previously, this can be shown to be

$$\log y = 0{\cdot}03x + 2{\cdot}569$$

Table XXXI

Calculation of the coefficient of variation of actual decadal values from hypothetical decadal values based on the logarithmic regression line

i	ii	iii	iv	v	vi	vii
(*a*)	(log *b*)	(hypothetical *b*)	(actual *b*)	(iv − iii)	$\left(\frac{\text{v}}{\text{iii}}.100\%\right)$	(vi²)
1	2·599	397·2	390	− 7·2	−1·81	3·28
2	2·629	425·6	435	+ 9·4	+2·21	4·88
3	2·659	456·0	475	+19·0	+4·17	17·40
4	2·689	488·7	480	− 8·7	−1·78	3·17
5	2·719	523·6	500	−23·6	−4·50	20·25
6	2·749	561·0	550	−11·0	−1·96	3·85
7	2·779	601·2	620	+18·8	+3·13	9·80
8	2·809	644·2	650	+ 5·8	+0·92	0·85
						7)63·48
						9·07

Coefficient of variation, or percentage standard deviation = $\sqrt{9{\cdot}07}$ = 3·01

From the actual and hypothetical values for the population set out in Table XXXI calculation proceeds very much along the lines of that for the χ^2 Test. Thus the difference between the observed (or actual) values and the expected (or hypothetical) values are first obtained (see also Fig. 48). It is then possible to work with these as *percentages* of the expected value and these differences (column v in Table XXXI) have been transferred to percentages of their respective expected values in column vi of the same table. This is necessary

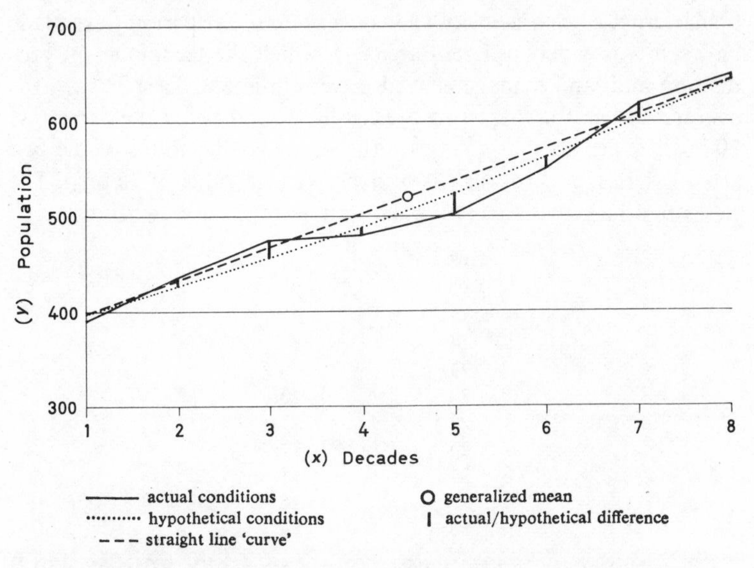

Figure 48. Deviation of observed population data from a logarithmic regression line

because the variability is being measured from the trend line, *not* from one mean value as in the case of the normal standard deviation. These percentage deviations from the trend are then squared, the values summed and divided by $(n - 1)$ rather than by (n), because of the small size of the sample. This gives, in percentage terms, the best estimate of the variance of the population values from the trend line. This value is here 9·07% and the square root of this, i.e. 3·01%, gives the best estimate of the percentage standard deviation (or the coefficient of variation) of these population data about the logarithmic regression line that represents the trend. This can be compared to the value of 17·4% given on p. 241, reflecting this variability *plus* the overall trend. This small value is the result of those factors that are *not* incorporated in the trend line. Such values as these allow the comparison between different units to be made, in terms of the extent to which population changes in those units deviate from the hypothetical changes (here based on the logarithmic curve).

A further example, this time based on a straight-line regression, will emphasize the method again, and also allow of a comparison being made between two sets of data. Assume that, for some particular crop, comparisons of yields over a ten-year period were made between two widely different areas, in which the techniques of production and land management also were different. Despite this, the average values for these two areas were found to be the same (i.e. 20 bushels per acre) as also were the standard deviations of the two sets of data (i.e. 3·16 bushels or a percentage value of 15·8%). The yields for these two areas for each year were as set out below.

Area I	Area II
16	24
17	23
24	24
23	20
20	20
20	23
23	17
24	17
17	16
16	16

Clearly the year-to-year variations are markedly different, and it would, of course, be possible to compare these by means of a correlation coefficient. Equally, however, it would be possible to calculate a straight regression line for each set of data, to see whether any overall trend occured. By applying the methods outlined in Table XXVI, the regression formulae would be:

Area I $y = +20$ } where $x =$ the time-scale
Area II $y = 25{\cdot}4 - 0{\cdot}982\underline{x}$ } and $y =$ the crop yield

Thus in Area I there is no overall trend at all (see Fig. 49), so that the percentage variability value of 15·8% reflects variability about one constant value, i.e. it illustrates the influence of such factors as annual variations in climate or seed quality, etc. The actual cause can-

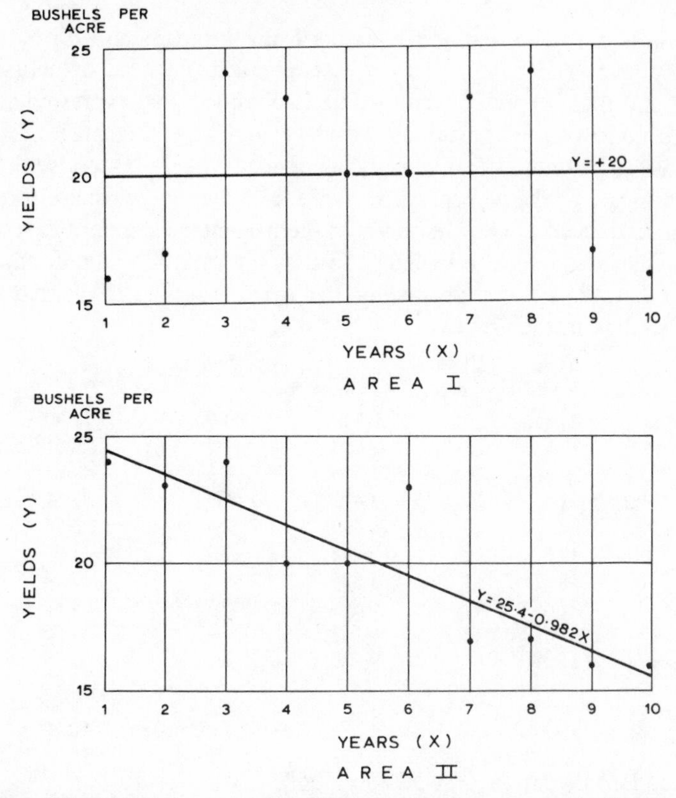

Figure 49. Straight-line regressions for crop yields for two areas for the same ten years

not, of course, be obtained without further analysis of possible causative factors. In Area II, however, a marked overall trend can be seen (Fig. 49) which consists of a fall in yields as time passes. Thus there is some dominant factor at work leading to decreasing yields (e.g. decreasing soil fertility because of agricultural practices; a progressive deterioration in climate), and this factor is hidden in the overall variability of 15·8%. In such a case it is useful to be able to separate the variability that results from factors other than those that induce declining yields, and this can be done by the present technique of assessing variability from the trend.

The necessary calculations for this are set out below. These consist of first obtaining for each year the values from the trend, either from the graph in Fig. 49 or by calculating from the regression formula given above. Then the difference between each observed or actual value and these expected trend values is obtained, and expressed as a percentage of the appropriate trend value. These percentages are squared and summed, this value being divided by $(n - 1)$ to give the best estimate of the percentage variance, and finally the square root obtained to yield the best estimate of the percentage standard deviation. This is now seen to be 7·6%, so that the variability due to factors other than those producing the overall decline is markedly smaller than in the first case.

Year	Yield (trend)	Yield (actual)	Difference	Difference %	$\%^2$
1	24·418	24	0·418	1·7	2·89
2	23·436	23	0·436	1·9	3·61
3	22·454	24	1·546	6·9	47·61
4	21·472	20	1·472	6·9	47·61
5	20·490	20	0·490	2·4	5·76
6	19·508	23	3·492	17·9	321·31
7	18·526	17	1·526	8·2	67·24
8	17·544	17	0·544	3·1	9·61
9	16·562	16	0·562	3·4	11·56
10	15·580	16	0·420	2·7	7·29
					9)524·49
				% variance =	58·28

% standard deviation = $\sqrt{58 \cdot 28} = 7 \cdot 6\%$

It has no doubt occurred to the reader that this whole process is akin to the calculation of the residual variance and the standard error of the estimate considered on pp. 214–216. It is, in fact, exactly the same concept involved, except that this is applied to a regression line for a single variable with the values occurring at regular intervals. It would be possible instead to calculate a correlation coefficient between the yield values for Area II and the years involved (simply numbered 1 to 10). This gives a coefficient of −0·892, and from this the regression equation, obtained by any of the methods given in Chapter 12, is once again

$$y = 25 \cdot 4 - 0 \cdot 982\, x$$

If, from these values, the standard error of the estimate of y were to be obtained (see p. 214), using the best estimate of the standard deviation of the yield data for comparability with the calculations on p. 245, it would be found that this standard estimate is 1·51 bushels per acre. As the average value for yield is 20 bushels per acre, this means that the percentage relationship is again 7·6%, as obtained previously on p. 245. In other words, the method outlined on pp. 240–245 for estimating deviations from a trend line, also provides a technique by which confidence limits can be obtained for a regression line of one variable against regular recording intervals (see p. 224).

Rhythmic Fluctuations

In the graph of crop yields for Area I (Fig. 49) it can be seen that there appears to be some semi-rhythmic fluctuation in the values, this rhythm being so regular that it can be smoothed out into a trend line that displays no overall change. Fluctuations that follow, regularly or irregularly, a semi-rhythmic pattern require yet more advanced techniques for their definition and elucidation. Many phenomena, whose data cannot be represented by *one* trend (whether straight-line or curvilinear), may nevertheless correspond to the overlapping of several dissimilar rhythms or waves. To define these requires some ability in harmonic analysis, a technique that must remain beyond the scope of an introductory book such as this. The reader is referred to more advanced statistical texts if some proficiency in harmonic analysis is required for research purposes. If, in a long series of data, there are clearly several distinct trends, it is, however, always possible to compute the regression for each of these periods separately. This will provide a closer approximation to the trend in such cases than will the reliance on one simple regression line that groups several smaller but distinctive trends together. Thus, although this must remain a 'second best' as compared to harmonic analysis, being both generalized and in part subjective, it does provide a simple method of making a first approximation to a series of regression lines that will show changes in trends from one period to another.

CHAPTER 14

SCOPE FOR THE FUTURE

Throughout all the preceding chapters the conscious aim has been to present, as simply as possible, the basic elements of a wide range of statistical techniques. All of these techniques are standard ones and have been widely applied in many fields of study, where they form an essential tool in the analysis of numerical data. Without such techniques these fields of study would not have progressed as steadily and effectively as they have done. They have allowed the conclusions of experimental or observational studies to be presented in a form that is common to all fields that attempt to express their results quantitatively. Furthermore, the use of these methods helps to reduce the element of subjective judgment in so many ways, thus ensuring that from the same set of data different workers will arrive at roughly the same conclusion. In this way it is possible for studies to be repeated so that cross-checking of results can be effected, while it also means that the mental reasoning by which a certain conclusion is arrived at is clearly apparent to all later workers. The gains thus include greater clarity, objectivity, orderliness and precision.

This is *not* to argue, however, that only conclusions based on statistical methods are of any validity. While some problems lend themselves to analysis by such methods, being concerned with quantitative data of one sort or another, others can only be resolved by personal assessment based on experience, ability and the proper understanding of the phenomena under study. Even in these cases, however, it is often true that such personal assessments can be considerably assisted and facilitated by the use of statistical analysis at one or more stages in the study. Equally, experience, ability and understanding are essential before any study based on statistical methods can be expected to yield valuable and relevant results. In other words, statistical techniques are simply a series of special tools which can be of as much assistance in the study of geographical problems as they have proved to be in problems of the pure sciences, other field sciences and the social sciences. This does not mean that they will be of equal value in all problems that confront geographers, any more than palaeography, pollen analysis or surveying are always relevant to any particular problem. It does mean, however, that whenever statistical analysis *is* relevant and *is* required, then the geographer should use such techniques to the fullest extent that is necessary to ensure a satisfactory solution of his problem.

The use of such techniques necessarily implies a proper understanding of them. Only in this way can a sound choice be made between differing methods, the data be organized in a suitable form, and the correct interpretation be made of the results. This is just as important if, as often in more complex problems, the geographer must seek guidance from a professional statistician, otherwise the most refined techniques may lead to erroneous conclusions through mis-interpretation. For such an understanding the simple concepts and techniques presented in this book are essential. In many studies these simpler techniques will be all that is required, but even if more advanced and complex techniques are needed for particular problems most of them will be found to be related to these simple concepts.

In an ever-growing proportion of current geographical research, however, recourse is being made to these more complex techniques of statistical analysis, especially to the many methods of multivariate analysis. In the problems and methods presented in this book, attention has almost invariably been focused on only *one* possible causative factor at a time, but of greater relevance in geographical research is the realization that any pattern or distribution being studied is the result of a *complex* of variable factors, or at least of several major factors within that complex. It is therefore necessary to accept that in any problem there are a whole host of partial causes contributing to the ultimate reality. It is, of course, possible to look at the impact of each of these in turn, using methods such as those outlined in this book, but this ignores two fundamental characteristics —first, that these several variables are likely to be interrelated between themselves so that the apparent impact of any one of them may in fact be due partially or wholly to the presence of some other factor; second, that apart from the impact of each variable separately, there may be an additional impact resulting from the co-existence of two or more variables. Clearly, to separate these impacts, and to assess their relative importance, gives a far more meaningful explanation of the problem under study than any judgment based on 'feel' or simple guesswork, no matter how intelligently carried out.

One such group of techniques represents expansion of the analysis of variance theme outlined in Chapter 9, especially various types of factorial design incorporating the Yates' 2^n procedure and also the Latin Square technique in sampling in relation to it. Another group of techniques utilizes the basic concepts of correlation and regression presented in Chapters 11, 12 and 13 applying them to cases with a number of independent variables by means of multiple and partial correlation coefficients and multiple regression equations. The analysis of the resulting residuals, and the construction of trend surfaces of varying degrees of complexity, develop from such studies. Correlation coefficients also form the initial basis for the synthesizing techniques of factor analysis and principal components analysis, which are of rapidly expanding application, and equally of the simpler method of linkage analysis. On somewhat different lines, the construction of mathematical models based on advanced probability theory (Chapter 5) and game theory concepts, either to summarize the complexity of reality or to provide criteria against which to test reality, is of growing importance and significance. Moreover, in all these studies, problems of sampling design (Chapters 6 and 7) have to be faced and solved, especially when it is upon field data that the investigation is based. These, and other more advanced techniques, require a thorough appreciation and understanding of statistical reasoning and methodology. They are fraught with difficulties and dangers for the ill-prepared, not only in terms of computation but even more so in terms of interpretation and application; to set against this, however, they present potentialities which are far from fully explored in a geographical context, and which are still not fully appreciated by many geographers. These potentialities amply repay the individual who equips himself to proceed from the simple basic concepts outlined in the foregoing pages to a facility with, and a control of, these more advanced methods of statistical analysis.

The major practical problem in applying these or other techniques is likely to be related to the time consumed in making the necessary calculations. While it is true that practice greatly increases speed (and one hopes accuracy, too), some mechanical means of assistance is essential. Facility with a slide rule is almost a *sine qua non* if numerous calculations are being made, and with this the individual student can cope with quite substantial calculations in a reasonable space of time. For larger problems, however, especially when the body of data is considerable, a mechanical calculating machine is almost indispensable. Desk models, operated manually, electrically or electronically, can allow of great quantities of data being processed with perfect accuracy and relatively little strain. In view of the wide range of geographical problems that can be approached, at least in part, via statistical analysis, it would seem more useful for geography students to be proficient with a calculating machine than with a theodolite or meteorological instruments, with their more limited application! At research level, of course, it is now possible to employ electronic computers to effect lengthy and involved calculations exceedingly rapidly, although the time taken up in the initial data tape preparation and programming should always be borne in mind. Nevertheless the existence of such computers in most universities, as well as in research and business establishments, now opens up the possibility of tackling fairly quickly problems of a magnitude that formerly could not have been contemplated. The large-scale study need no longer be either excessively generalized or else a lifetime's task, but rather it may be a major project lasting a period of two to three years. The use of such computers necessarily requires training and practice, but with assistance from those directly concerned with them this is a feasible proposition. It does mean, however, that at least the simpler techniques of statistical analysis must be known and understood not only by those carrying out such studies, but also by all geographers who are going to use or interpret the results obtained in this way.

Finally it must be stressed that facility with any of these techniques, whether they be simple or complex ones, will only come by continued use and practice. This is especially true for those geographers—and they are the majority, unfortunately—who have used mathematical methods for little more than everyday purposes since the age of sixteen. Once a certain familiarity has been established with these methods, however, the possible uses of them become increasingly apparent. The problems presented in this book represent but a small selection of very simple types of problem that could have been considered, and these techniques are expanding into all those aspects of geography where they have any relevance. Provided that these methods of analysis are then kept in their proper place, i.e. as a tool by which geographical studies can be furthered, and not as an end in themselves, they can provide a positive contribution to the expansion and value of geography as a whole.

A SELECTIVE BIBLIOGRAPHY

The following books represent a brief cross-section of the large literature on statistical methods and their application. Section A comprises books that are concerned with the broad field as a whole; the later sections contain publications more specifically concerned with one or other particular aspect of these techniques. Such a distinction cannot, of course, be a clear-cut one, and there is often a measure of overlap between the various sections. Moreover, each section includes works of two kinds—those that are primarily concerned with the method under review, and which are therefore not necessarily geographical in any real sense; and those that exemplify the application of such techniques in the field of geography or some kindred discipline. The selection of publications is essentially a personal one; it focuses on the more readily accessible publications in the English language, but even within these limits it does not attempt to be comprehensive. It is reasonably representative, however, and many of the works listed include bibliographies which cover more fully the publications available in their particular fields. It is hoped and intended that by reference to the books and papers listed below, it will be possible for the reader both to expand the examples of the simple methods that have been outlined in this book, and also to consider more advanced techniques some of which have been referred to in passing in the previous pages.

A. *General texts*

ALLEN, R. G. D., *Statistics for economists*, Hutchinson University Library, 1957.

BALCHIN, W. G. V., ed., *Geography: an outline for the intending student*, Routledge and Kegan Paul, 1970.

BERRY, B. J. L. and MARBLE D. F., eds., *Spatial analysis: a reader in statistical geography*, Prentice-Hall, 1968.

BRANFIELD, D. J. R. and BELL, H. W. *Matrices and their applications*, Macmillan, 1970.

BROOKS, C. E. P. and CARRUTHERS, N., *Handbook of statistical methods in meteorology*, H.M.S.O., 1953.

BUNGE, W. W., *Theoretical geography*, Lund Studies in Geography, 1966.

BURTON, I., 'The quantitative revolution', *Canadian Geographer*, **7,** 151, 1963.

CHORLEY, R. J., 'The application of statistical methods to geomorphology', *Essays in geomorphology*, ed. G. H. Dury, 275,1966.

CHORLEY, R. J. and HAGGETT, P., eds., *Models in geography*, Methuen, 1967.

COLE, J. P. and KING, C. A. M., *Quantitative geography*, Wiley, 1968.

CONRAD, V. and POLLAK, L. W., *Methods in climatology*, Oxford University Press, 1962.

FRENCH, H. M. and RACINE, J. B., eds., *Quantitative and qualitative geography*, University of Ottawa Press, 1971.

GARRISON, W. L., 'The applicability of statistical inference to geographical research', *Geographical Review*, **46,** 427, 1956.

GREGORY, S., 'The role of models and quantitative techniques in teaching—attitudes, opinions and prejudices', *Geography*, **54,** 5, 1969.

GREIG-SMITH, P., *Quantitative plant ecology*, Butterworth, 1964.

HÄGERSTRAND, T., 'The computer and the geographer', *Transactions of the Institute of British Geographers*, **42,** 1, 1967.

HAGGETT, P., *Locational analysis in human geography*, Arnold, 1965.

HARVEY, D. W., *Explanation in geography*, Arnold, 1969.

KERSHAW, K. A., *Quantitative and dynamic ecology*, Arnold, 1964.

KING, L. J., *Statistical analysis in geography*, Prentice-Hall, 1969.

—— 'The analysis of spatial form and its relation to geographic theory', *Annals of the Association of American Geographers*, **59,** 573, 1969.

KRUMBEIN, W. C. and GRAYBILL, F. A., *An introduction to statistical models in geology*, McGraw-Hill, 1965.

LAVALLE, P., MCCONNELL, H., and BROWN, R. G., 'Certain aspects of the expansion of quantitative methodology in American Geography', *Annals of the Association of American Geographers*, **57,** 423, 1967.

LEOPOLD, L. B., WOLMAN, M. G., and MILLER, J. P., *Fluvial processes in geomorphology*, Freeman, 1964.

LINDLEY, D. V. and MILLER, J. C. P., *Cambridge elementary statistical tables*, Cambridge University Press, 1953.

MILLER, R. L. and KAHN, J. S., *Statistical analysis in the geological sciences*, Wiley, 1962.

MOORE, P. G., *Principles of statistical techniques*, Cambridge University Press, 1958.

MORONEY, M. J., *Facts from figures*, Pelican, 1964.

PANOFSKY, H. A. and BRIER, G. W., *Some applications of statistics to meteorology*, Pa. State University College of Mineral Industries, 1958.

PATERSON, D. D., *Statistical techniques in agricultural research*, McGraw-Hill, 1939.

REYNOLDS, R. B., 'Statistical methods in geographic research', *Geographical Review*, **46,** 129, 1956.

RIDER, P. R., *Introduction to modern statistical methods*, Wiley, 1939.

TIPPETT, L. H. C., *Methods of statistics*, Benn, 1952.

TIPPETT, L. H. C., *Statistics*, 2nd edn. Oxford University Press, Home University Library, 1956.

WALKER, H. M., *Mathematics essential for elementary statistics*, Holt, 1951.

YULE, G. U. and KENDALL, M. G., *An introduction to the theory of statistics*, Griffin, 1958.

B. *Sampling design*

BIRCH, J. W., 'A note on the sample-farm survey and its use as a basis for generalised mapping', *Economic Geography*, **36,** 254, 1960.

HAGGETT, P. and BOARD, C., 'Rotational and parallel traverses in the rapid integration of geographic areas', *Annals of the Association of American Geographers*, **54,** 406, 1964.

HANSEN, M. M., HURWITZ, W. N., and MADOW, W. G., *Sample survey methods and theory*, Wiley, 1953.

HOLMES, J., 'Problems in location sampling', *Annals of the Association of American Geographers*, **57,** 757, 1967.

KRUMBEIN, W. C., 'Experimental design in the earth sciences', *Transactions of the American Geophysical Union*, **36,** 1, 1954.

KRUMBEIN, W. C., 'The geological population as a framework for analysing numerical data in geology', *Liverpool and Manchester Geological Journal*, **2,** 341, 1960.

KRUMBEIN, W. C. and MILLER, R. L., 'Design of experiments for statistical analysis of geological data', *Journal of Geology*, **61,** 510, 1953.

SAMPFORD, M. R., *An introduction to sampling theory with applications to agriculture*, Oliver and Boyd, 1962.

WOOD, W. F., 'The use of stratified random samples in land use study', *Annals of the Association of American Geographers*, **45**, 350, 1955.

YATES, F., *Sampling methods for censuses and surveys*, Griffin, 1953.

C. *Significance testing*

ARMSTRONG, R. W., 'Standardized class intervals and rate computation in statistical maps of mortality', *Annals of the Association of American Geographers*, **59**, 382, 1969.

BARRETT, E. C., 'Local variations in rainfall trends in the Manchester region', *Transactions of the Institute of British Geographers*, **35**, 55, 1964.

CROWE, P. R., 'The analysis of rainfall probability: a graphical method and its application to European data', *Scottish Geographical Magazine*, **49**, 73, 1933.

CROWE, P. R., 'The rainfall regime of the Western Plains', *Geographical Review*, **26**, 463, 1936.

GREGORY, S., 'Regional variations in the trend of annual rainfall over the British Isles', *Geographical Journal*, **122**, 346, 1956.

SCOTT, P., 'Areal variations in the class structure of the central place hierarchy', *Australian Geographical Studies*, **2**, 73, 1964.

SIEGEL, S., *Nonparametric statistics for the behavioral sciences*, McGraw-Hill, 1956.

ZOBLER, L., 'Statistical testing of regional boundaries', *Annals of the Association of American Geographers*, **47**, 83, 1957; and replies by MACKAY, J. R. and BERRY, B. J. L., **49** p. 89, 1959.

D. *Analysis of variance and its derivatives*

DUNCAN, O. D., CUZZORT, R. P., and DUNCAN, B., *Statistical geography: problems in analysing areal data*, Collier-Macmillan, 1961.

HAGGETT, P., 'Regional and local components in the distribution of forested areas in S.E. Brazil: a multi-variate approach', *Geographical Journal*, **130**, 365, 1964.

HILL, A.R., 'An experimental test of the field technique of till macrofabric analysis', *Transactions of the Institute of British Geographers*, **45**, 93, 1968.

KARIEL, H. G., 'Selected factors areally associated with population growth due to net migration', *Annals of the Association of American Geographers*, **53**, 210, 1963.

ROBINSON, A. H. and BRYSON, R. A., 'A method of describing quantitatively the correspondence of geographical distributions', *Annals of the Association of American Geographers*, **47**, 379, 1957.

SNEDECOR, G. W., *Statistical methods applied to experiments in agriculture and biology*, Iowa State College Press, 1946.

E. *Correlation, regression and trends*

BAGGALEY, A. R., *Intermediate correlation methods*, Wiley, 1964.

BRUSH, J. E., 'Spatial patterns of population in Indian cities', *Geographical Review*, **58**, 362, 1968.

CHORLEY, R. J. and HAGGETT, P., 'Trend surface mapping in geographical research', *Transactions of the Institute of British Geographers*, **37**, 47, 1965.

COULSON, M. R. C., 'The distribution of population age structures in Kansas City', *Annals of the Association of American Geographers*, **58**, 155, 1968.

EZEKIEL, M. and FOX, K. A., *Methods of correlation and regression analysis*, Wiley, 1959.

GEDDES, A., 'The population of India: variability of change as a regional demographic index', *Geographical Review*, **32**, 562, 1942.

GEDDES, A., 'Variability in change of population in the United States and Canada, 1900–1951', *Geographical Review*, **44**, 88, 1954.

GREGG, J. V., HOSSELL, C. H. and RICHARDS, J. T., *Mathematical trend curves: an aid to forecasting*, Oliver and Boyd, 1964.

GREGORY, S., *Rainfall over Sierra Leone* (Section 2), Liverpool University Department of Geography, No. 2, 1965.

HART, J. F. and SALISBURY, N. E., 'Population change in Middle Western villages: a statistical approach', *Annals of the Association of American Geographers*, **55**, 140, 1965.

HURST, M. E. E., 'An approach to the study of non-residential land use traffic generation', *Annals of the Association of American Geographers*, **60**, 153, 1970.

KENDALL, M. G., *Contributions to the study of oscillatory time-series*, Cambridge, 1946.

KENYON, J. B., 'On the relationships between central function and size of place', *Annals of the Association of American Geographers*, **57**, 736, 1967.

KING, C. A. M., 'Trend surface analysis of Central Pennine erosion surfaces', *Transactions of the Institute of British Geographers*, **47**, 47, 1969.

KING, L. J., 'A multivariate analysis of the spacing of urban settlements in the United States', *Annals of the Association of American Geographers*, **50**, 157, 1960.

KRUMBEIN, W. C., 'Trend surface analysis of contour-type maps with irregular control-point spacing', *Journal of Geophysical Research*, **64**, 823, 1959.

LAVALLE, P., 'Some aspects of linear karst depression development in South Central Kentucky', *Annals of the Association of American Geographers*, **57**, 49, 1967.

ROBINSON, A. H., 'Mapping the correspondence of isarithmic maps', *Annals of the Association of American Geographers*, **52**, 414, 1962.

ROBINSON, G. and FAIRBAIRN, K. J, 'An application of trend-suface mapping to the distribution of residuals from a regression', *Annals of the Association of American Geographers*, **59**, 158, 1969.

ROBINSON, A. H., LINDBERG, J. B. and BRINKMAN, L. W., 'A correlation and regression analysis applied to rural farm population densities in the Great Plains', *Annals of the Association of American Geographers*, **51**, 211, 1961.

RUSSWURM, L. H., 'The central business district retail sales mix, 1948–1958', *Annals of the Association of American Geographers*, **54**, 524, 1964.

SHUE TUCK WONG, 'A multivariate statistical model for predicting mean annual floods in New England', *Annals of the Association of American Geographers*, **53**, 298, 1963.

WOLPERT, J., 'The decision process in spatial context', *Annals of the Association of American Geographers*, **54**, 537, 1964.

—— 'Vertical displacement of shorelines in Highland Britain', *Transactions of the Institute of British Geographers*, **39**, 1966.

F. *Factor analysis and similar techniques*

ADCOCK, C. J., *Factorial analysis for non-mathematicians*, Cambridge, 1954.

ABIODUN, J. O., 'Urban hierarchy in a developing country', *Economic Geography*, **43**, 347, 1967.

BROWN, S. E. and TROTT, C. E., 'Grouping tendencies in an economic regionalization of Poland', *Annals of the Association of American Geographers*, **58**, 327, 1968.

CATTELL, R. B., 'Factor analysis: an introduction to essentials', *Biometrics*, **21**, 190 and 405, 1965.

COX, K. R., 'Suburbia and voting behavior in the London metropolitan area', *Annals of the Association of American Geographers*, **58**, 111, 1968.

GENTILCORE, R. L., 'Reclamation in the Agro Pontino, Italy', *Geographical Review*, **60**, 301, 1970.

GINSBURG, N., *Atlas of Economic development* (Part VIII—'A statistical analysis', by BERRY, B. J. L.), University of Chicago Press, 1961.

GOULD, P. R., 'On the geographical interpretation of eigenvalues', *Transactions of the Institute of British Geographers*, **42**, 53, 1967.

GREGORY, S., *Rainfall over Sierra Leone* (Section 3), University of Liverpool Department of Geography, No. 2, 1965.

HARMAN, H. H., *Modern factor analysis*, University of Chicago Press, 1967.

HENSHALL, J. D., 'The demographic factor in the structure of agriculture in Barbados', *Transactions of the Institute of British Geographers*, **38**, 183, 1966.

JOHNSTON, R. J., 'Choice in classification; the subjectivity of objective methods', *Annals of the Association of American Geographers*, **58**, 575, 1968.

JOWETT, G. H., 'Factor analysis', *Applied Statistics*, **7**, 114, 1958.

KENDALL, M. G., 'The geographical distribution of crop productivity in England', *Journal of the Royal Statistical Society*, **102**, 21, 1939.

KING, L. J., 'Discriminatory analysis of urban growth patterns in Ontario and Quebec, 1951-1961', *Annals of the Association of American Geographers*, **57**, 566, 1967.

LAWLEY, D. N. and MAXWELL, A. E., *Factor analysis as a statistical method*, Butterworth, 1963.

MATHER, P. M. and DOORNKAMP, J. C., 'Multivariate analysis in geography with particular reference to drainage basin morphometry', *Transactions of the Institute of British Geographers*, **51**, 163, 1970.

MAXWELL, A. E., 'Recent trends in factor analysis', *Journal of the Royal Statistical Society, Series A (General)*, **124**, 49, 1961.

MAYFIELD, R. C., 'The range of a central good in the Indian Punjab', *Annals of the Association of American Geographers*, **53**, 38, 1963.

MOSER, C. A. and SCOTT, W., *British towns: a statistical study of their social and economic differences*, Oliver and Boyd, 1961.

PERRY, A. H., 'Filtering climatic anomaly fields using principal component analysis', *Transactions of the Institute of British Geographers*, **50**, 55, 1970.

POCOCK, D. C. D. and WISHART, D., 'Methods of deriving multi-factor uniform regions', *Transactions of the Institute of British Geographers*, **47**, 73, 1969.

G. *Probability studies and models*

BERRY, B. J. L., 'Approaches to regional analysis: a synthesis', *Annals of the Association of American Geographers*, **54**, 2, 1964.

BERRY, B. J. L., *Geography of market centres and retail distribution*, Prentice-Hall, 1967.

CHORLEY, R. J., 'Geomorphology and general systems theory', *U.S. Geological Survey, Professional Paper*, 500B, 1962.

CHORLEY, R. J., 'Geography and analogue theory', *Annals of the Association of American Geographers*, **54**, 127, 1964.

CHORLEY, R. J. and HAGGETT, P., eds., *Models in geography: the Madingley lectures for 1965*, Methuen, 1967.

CHORLEY, R. J. and HAGGETT, P., *Network analysis*, Arnold, 1969.

CLARK, W. A. V. and RUSHTON, G., 'Models of intra-urban consumer behavior and their implication for central place theory', *Economic Geography*, **46**, 486, 1970.

CURRY, L., 'Climatic change as a random series', *Annals of the Association of American Geographers*, **52**, 21, 1962.

CURRY, L., 'The random spatial economy: an exploration in settlement theory', *Annals of the Association of American Geographers*, **54**, 138, 1964.

GETIS, A., 'Temporal land use pattern analysis with the use of nearest neighbour and quadrat methods', *Annals of the Association of American Geographers*, **54**, 391, 1964.

GOULD, P. R., 'Man against his environment: a game theoretic framework', *Annals of the Association of American Geographers*, **53,** 290, 1963.

GREGORY, S., 'Rainfall reliability', in *Environment and land use in in Africa*, eds., M. F. Thomas and G. W. Whttington, Methuen, p. 57, 1969.

HARVEY, D. W., 'Geographical processes and the analysis of point patterns', *Transactions of the Institute of British Geographers*, **40,** 81, 1966.

HARVEY, D. W., 'Some methodological problems in the use of the Neyman Type A and the Negative Binomial probability distributions for the analysis of spatial point patterns', *Transactions of the Institute of British Geographers*, **44,** 85, 1968.

HODGES, J. L., and LEHMANN, E. L., *Basic concepts of probability and statistics*, Holden-Day, 1964.

ISARD, W., *Location and space economy*, M.I.T. Press, 1962.

ISARD, W., *et al., Methods of regional analysis: an introduction to regional science*, Wiley, 1960.

JANELLE, D. G., 'Spatial reorganization: a model and concept', *Annals of the Association of American Geographers*, **59,** 348, 1969.

LUCKERMAN, F. and PORTER, P. W., 'Gravity and potential models in economic geography', *Annals of the Association of American Geographers*, **50,** 493, 1960.

MANNING, H. L., 'The statistical assessment of rainfall probability and its application in Uganda agriculture', *Proceedings of the Royal Society, Series B*, **144,** 460, 1956.

MORRILL, R. L., 'The development of spatial distributions of towns in Sweden: an historical predictive approach', *Annals of the Association of American Geographers*, **53,** 1, 1963.

PRED, A. R. and KIBEL, B. M., 'An application of gaming simulation to a general model of economic locational processes', *Economic Geography*, **46,** 136, 1970.

WHITTLE, P., *Probability*, Penguin Books, 1970.

Formulae Index

FORMULAE INDEX

The variation of type indicate the following:
6 (chapter); 6 (page); *6* (figure); VI (table)

General Index

GENERAL INDEX

The variation of type indicate the following:
6 (chapter); 6 (page); *6* (figure); VI (table)